BULK COLLECTION OF
SIGNALS INTELLIGENCE
TECHNICAL OPTIONS

Committee on Responding to Section 5(d) of
Presidential Policy Directive 28:
The Feasibility of Software to Provide Alternatives to
Bulk Signals Intelligence Collection

Computer Science and Telecommunications Board

Division on Engineering and Physical Sciences

NATIONAL RESEARCH COUNCIL
OF THE NATIONAL ACADEMIES

THE NATIONAL ACADEMIES PRESS
Washington, D.C.
www.nap.edu

THE NATIONAL ACADEMIES PRESS   500 Fifth Street, NW   Washington, DC 20001

NOTICE: The project that is the subject of this report was approved by the Governing Board of the National Research Council, whose members are drawn from the councils of the National Academy of Sciences, the National Academy of Engineering, and the Institute of Medicine. The members of the committee responsible for the report were chosen for their special competences and with regard for appropriate balance.

Support for this project was provided by the Office of the Director for National Intelligence, Contract Number 2014-14041100003-001. Any opinions, findings, conclusions, or recommendations expressed in this publication are those of the author(s) and do not necessarily reflect the views of the organizations or agencies that provided support for the project.

International Standard Book Number 13:   978-0-309-32520-2
International Standard Book Number 10:   0-309-32520-X
Library of Congress Control Number: 2015933164

This report is available from

Computer Science and Telecommunications Board
National Research Council
500 Fifth Street, NW
Washington, DC 20001

Additional copies of this report are available from the National Academies Press, 500 Fifth Street, NW, Keck 360, Washington, DC 20001; (800) 624-6242 or (202) 334-3313; http://www.nap.edu.

Copyright 2015 by the National Academy of Sciences. All rights reserved.

Printed in the United States of America

THE NATIONAL ACADEMIES
Advisers to the Nation on Science, Engineering, and Medicine

The **National Academy of Sciences** is a private, nonprofit, self-perpetuating society of distinguished scholars engaged in scientific and engineering research, dedicated to the furtherance of science and technology and to their use for the general welfare. Upon the authority of the charter granted to it by the Congress in 1863, the Academy has a mandate that requires it to advise the federal government on scientific and technical matters. Dr. Ralph J. Cicerone is president of the National Academy of Sciences.

The **National Academy of Engineering** was established in 1964, under the charter of the National Academy of Sciences, as a parallel organization of outstanding engineers. It is autonomous in its administration and in the selection of its members, sharing with the National Academy of Sciences the responsibility for advising the federal government. The National Academy of Engineering also sponsors engineering programs aimed at meeting national needs, encourages education and research, and recognizes the superior achievements of engineers. Dr. C. D. Mote, Jr., is president of the National Academy of Engineering.

The **Institute of Medicine** was established in 1970 by the National Academy of Sciences to secure the services of eminent members of appropriate professions in the examination of policy matters pertaining to the health of the public. The Institute acts under the responsibility given to the National Academy of Sciences by its congressional charter to be an adviser to the federal government and, upon its own initiative, to identify issues of medical care, research, and education. Dr. Victor J. Dzau is president of the Institute of Medicine.

The **National Research Council** was organized by the National Academy of Sciences in 1916 to associate the broad community of science and technology with the Academy's purposes of furthering knowledge and advising the federal government. Functioning in accordance with general policies determined by the Academy, the Council has become the principal operating agency of both the National Academy of Sciences and the National Academy of Engineering in providing services to the government, the public, and the scientific and engineering communities. The Council is administered jointly by both Academies and the Institute of Medicine. Dr. Ralph J. Cicerone and Dr. C. D. Mote, Jr., are chair and vice chair, respectively, of the National Research Council.

www.national-academies.org

COMMITTEE ON RESPONDING TO SECTION 5(D) OF PRESIDENTIAL POLICY DIRECTIVE 28: THE FEASIBILITY OF SOFTWARE TO PROVIDE ALTERNATIVES TO BULK SIGNALS INTELLIGENCE COLLECTION

ROBERT F. SPROULL, University of Massachusetts, Amherst, *Chair*
FREDERICK R. CHANG, Southern Methodist University
WILLIAM H. DUMOUCHEL, Oracle Health Sciences
MICHAEL KEARNS, University of Pennsylvania
BUTLER W. LAMPSON, Microsoft Corporation
SUSAN LANDAU, Worcester Polytechnic Institute
MICHAEL E. LEITER, Leidos
ELIZABETH RINDSKOPF PARKER, University of the Pacific, McGeorge School of Law
PETER J. WEINBERGER, Google, Inc.

Staff

ALAN SHAW, Air Force Studies Board, *Study Director*
HERBERT S. LIN, Chief Scientist, CSTB
JON EISENBERG, Director, CSTB
ERIC WHITAKER, Senior Program Assistant, CSTB

COMPUTER SCIENCE AND TELECOMMUNICATIONS BOARD

ROBERT F. SPROULL, University of Massachusetts, Amherst, *Chair*
LUIZ ANDRÉ BARROSO, Google, Inc.
STEVEN M. BELLOVIN, Columbia University
ROBERT F. BRAMMER, Brammer Technology, LLC
EDWARD FRANK, Brilliant Lime and Cloud Parity
SEYMOUR E. GOODMAN, Georgia Institute of Technology
LAURA HAAS, IBM Corporation
MARK HOROWITZ, Stanford University
MICHAEL KEARNS, University of Pennsylvania
ROBERT KRAUT, Carnegie Mellon University
SUSAN LANDAU, Worcester Polytechnic Institute
PETER LEE, Microsoft Corporation
DAVID E. LIDDLE, US Venture Partners
BARBARA LISKOV, Massachusetts Institute of Technology
JOHN STANKOVIC, University of Virginia
JOHN A. SWAINSON, Dell, Inc.
ERNEST J. WILSON, University of Southern California
KATHERINE YELICK, University of California, Berkeley

Staff

JON EISENBERG, Director
VIRGINIA BACON TALATI, Program Officer
SHENAE BRADLEY, Senior Program Assistant
RENEE HAWKINS, Financial and Administrative Manager
HERBERT S. LIN, Chief Scientist
LYNETTE I. MILLETT, Associate Director
ERIC WHITAKER, Senior Program Assistant

For more information on CSTB, see its Web site at http://www.cstb.org, write to CSTB, National Research Council, 500 Fifth Street, NW, Washington, DC 20001, call (202) 334-2605, or e-mail the CSTB at cstb@nas.edu.

Preface

In January 2014, the President addressed the nation and the broader global community to explain U.S. policy regarding the collection of foreign intelligence. Shortly thereafter, the White House released Presidential Policy Directive 28 (PPD-28), in which Section 5(d) requested the Director of National Intelligence (DNI) to "assess the feasibility of creating software that would allow the IC more easily to conduct targeted information acquisition [of signals intelligence] rather than bulk collection."[1]

The Office of the Director of National Intelligence (ODNI) then asked the National Academies to form a committee to study this question, and discussions led to the charge to the committee shown in Box P.1. Note that the charge does not request recommendations, and the analysis and conclusions of the Committee on Responding to Section 5(d) of Presidential Policy Directive 28: The Feasibility of Software to Provide Alternatives to Bulk Signals Intelligence Collection are made with this in mind.

The committee assembled for this study included individuals with expertise in national security law; counterterrorism operations; privacy and civil liberties as they relate to electronic communications; data mining; large-scale systems development; software development; Intelligence Community (IC) needs as they relate to research and development; and networking and social media. See Appendix C for biographical information.

---

[1] The White House, Presidential Policy Directive/PPD-28, "Signals Intelligence Activities," Office of the Press Secretary, January 17, 2014, http://www.whitehouse.gov/sites/default/files/docs/2014sigint_mem_ppd_rel.pdf.

> **BOX P.1**
> **The Charge to the Committee**
>
> A committee appointed by the National Research Council will assess "the feasibility of creating software that would allow the U.S. intelligence community more easily to conduct targeted information acquisition rather than bulk collection," as called for in section 5(d) of Presidential Policy Directive 28. To the extent possible, it will consider the efficacy, practicality, and privacy implications of alternative software architectures and uses of information technology, and explore tradeoffs among these aspects in the context of representative "use cases." The study will consider a broad array of communications modalities, e.g., phone, email, instant message, and so on. It will not address the legality or value of signals intelligence collection. The study will identify and assess options and alternatives but will not issue recommendations.
>
> Specifically, the committee will address the following:
>
> 1. What are a small set of representative use cases within which one can explore alternative software architectures and uses of information technology, and consider trade-offs?
> 2. What is the current state of the software technology to support targeted information acquisition? What are feasible and likely trajectories for future relevant software development; near, mid, and far term? What are possible technology alternatives to bulk collection in the context of the use cases?
> 3. What are relevant criteria or metrics for comparing bulk collection to targeted collection (e.g. effectiveness, response time, cost, efficacy, practicality, privacy impacts)?
> 4. What tradeoffs arise with the technology alternatives analyzed in the context of the use cases and criteria/metrics?
> 5. How might requirements for information collection be altered in light of this analysis?
> 6. What uncertainties are associated with the assumptions and analyses, and how might they affect the basis for decisions?

With 5 months from study inception to delivery, the study committee was not blessed with a luxury of time. The committee sought to be responsive to the context in which the report was requested. In general terms, the committee saw its mission as exploring whether technological software-based alternatives to bulk collection might be identified in order to retain, to the extent possible, current intelligence capabilities while intruding less on parties that are not of known or potential interest to the IC. The legal protections provided by the Fourth Amendment and legislation such as the Foreign Intelligence Surveillance Act distinguish between foreign and U.S. persons; a factor that informed the committee's thinking.

The technological focus of this report is *not* limited to the metadata of domestic telephone communications, even though most public controversy has been pointed in this direction. Nor is the legal environment presumed to be only that governed by Section 215 of the Foreign Intelligence Surveillance Act—the legal authority under which the collection of telephone metadata has occurred. This report addresses the question of alternatives to bulk collection, without regard to the specific authorities and restrictions that control the various types of bulk collection. The types of communications of potential interest include any type of electronic communication. In the committee's view, signals intelligence has come to embrace almost any data stored on an electronic device. In a future that contains the Internet of Things, the scope will be even greater.

Furthermore, the committee chose to interpret its technological mandate broadly by considering a variety of approaches to reducing the degree of intrusiveness into the affairs of parties that are not of interest for intelligence purposes. Broadly, these approaches include the following:

- Collecting and/or storing less information,
- Better protecting the information that is collected or stored against theft or compromise, and
- Rigorously enforcing the rules governing use of collected or stored information.

Following its charge, the committee tried to confine its attention to *technical* aspects of signals intelligence and to avoid straying into legal and policy matters as much as possible. Despite this focus, there are areas of overlap and interdependence. For example, the more complex the rules and regulations established by policy and law, the more difficult it is to use automation to enforce them.

The situation with respect to bulk collection was a moving target during the time the report was written. During the final several weeks when the committee was responding to reviewers' comments, the Senate considered the USA Freedom Act (S.2685); this bill would have changed the collection of bulk business records. Providing value in this report meant focusing on collection options and their implications, rather than more narrowly tailoring the discussion to what the law presently provides. Thus the committee did not attempt, for example, to discuss what the implications of the proposed legislation might be on collection.

ODNI requested an unclassified report, with a classified annex if necessary. Nothing learned in classified briefings changed the committee's view or provided information essential to understanding the most important points of this report. The committee thus produced an entirely unclassified report, with no classified annex. The committee believes this unclassified report suffices to answer its charge to the best of its ability. One consequence of this approach is that some details must be omitted to protect sources and methods that the IC rightly guards with care.

An unclassified report risks being overtaken by newly declassified material. As this report was being finalized, documents were being declassified by the IC (see http://icontherecord.tumblr.com/) and released as a result of Freedom of Information Act requests. As a result, numerous omissions are bound to appear in the report; these omissions are not expected to change the committee's fundamental arguments, although new information may change details along the way.

The committee met six times in person, with the first meeting in mid-June 2014, and held numerous conference calls. Open sessions during its meetings were devoted to briefings from outside parties, and closed sessions were devoted to committee deliberations.

ACKNOWLEDGMENTS

The complexity and classified aspects of the issues explored in this report meant that the committee had much to learn from its briefers. The committee is grateful to many parties for presentations on:

- *June 30-July 2, 2014.* Joel Brenner (Joel Brenner LLC, the Chertoff Group, and former Inspector General, National Security Agency [NSA]), Carmen Medina (Deloitte Consulting LLP and former Deputy Director for Intelligence, Central Intelligence Agency [CIA]), Mark Maybury (The MITRE Corporation), General Keith B. Alexander (retired), Chris Inglis (former Deputy Director, National Security Agency), Wesley Wilson (ODNI/National Counterterrorism Center), Robert Brose (ODNI), William Crowell (Alsop-Louie Partners), Stephanie O'Sullivan (ODNI), David Honey (ODNI), and Marjory Blumenthal (Office of Science and Technology Policy).
- *August 4-6, 2014.* Jeff Jonas (IBM), Mark Lowenthal (Intelligence and Security Academy), and Philip Mudd (New America Foundation, Mudd Management, and former Deputy Director, CIA Counterterrorism Center).
- *August 27-29, 2014.* David Grannis (Senate Select Committee on Intelligence) and Kate Martin (Center for National Security Studies).
- *September 8-10, 2014.* Alexander Joel (ODNI), J.C. Smart (Georgetown University), Peter Highnam (Intelligence Advanced Research Projects Activity), and members of the Privacy and Civil Liberties Oversight Board.

The committee requested but did not receive comments from the American Civil Liberties Union, the Electronic Frontier Foundation, and the Electronic Privacy Information Center.

The committee appreciates the support of David Honey (Assistant Deputy Director of National Intelligence for Science and Technology [ADDNI/S&T]), Steven D. Thompson (Senior S&T Advisor), John C. Granger (Senior Advisor to the ADDNI/S&T), and their colleagues from ODNI who helped make this study possible and the many officials of ODNI and NSA who briefed the committee or answered its questions. In addition, the committee acknowledges the intellectual contributions of its staff, Alan Shaw (Study Director, Air Force Studies Board), Herbert S. Lin (Chief Scientist, Computer Science and Telecommunications Board [CSTB]), and Jon Eisenberg (Director, CSTB); consultants Alex Gliksman (AGI Consulting, LLC), M. Anthony Fainberg (Institute for Defense Analyses), and Allan Friedman (George Washington University); and Eric Whitaker (Senior Program Assistant, CSTB), who provided administrative support.

THE COMMITTEE'S PERSPECTIVE ON ITS CHARGE

This report is part of the national discussion about the balance between the powers of government and the rights of the governed, as the government tries to carry out its constitutionally mandated responsibilities. As indicated above, the committee was asked a question about technology. Accordingly, this report emphasizes technology but also attends to the need for effective and trustworthy processes, even as more sophisticated technologies are developed. But neither technology nor process—alone or together—can guarantee the proper balance between collective and individual security.

Acknowledgment of Reviewers

This report has been reviewed in draft form by individuals chosen for their diverse perspectives and technical expertise, in accordance with procedures approved by the National Research Council's Report Review Committee. The purpose of this independent review is to provide candid and critical comments that will assist the institution in making its published report as sound as possible and to ensure that the report meets institutional standards for objectivity, evidence, and responsiveness to the study charge. The review comments and draft manuscript remain confidential to protect the integrity of the deliberative process. We wish to thank the following individuals for their review of this report:

Steven M. Bellovin, Columbia University,
Joel F. Brenner, Joel Brenner LLC,
Fred H. Cate, Indiana University,
George R. Cotter, Isologic, LLC,
William P. Crowell, Alsop Louie Partners,
Michael V. Hayden, Chertoff Group,
Raymond Jeanloz, University of California, Berkeley,
Anita K. Jones, University of Virginia,
Orin S. Kerr, George Washington University,
Peter Lee, Microsoft Research,
Kate Martin, Center for National Security Studies, and
Cynthia Storer, Coastal Carolina University.

Although the reviewers listed above have provided many constructive comments and suggestions, they were not asked to endorse the report's conclusions, nor did they see the final draft of the report before its release. The review of this report was overseen by Samuel H. Fuller, Analog Devices, Inc., and William H. Press, University of Texas, Austin. Appointed by the National Research Council, they were responsible for making certain that an independent examination of this report was carried out in accordance with institutional procedures and that all review comments were carefully considered. Responsibility for the final content of this report rests entirely with the authoring committee and the institution.

Contents

SUMMARY 1

1 INTRODUCTION AND BACKGROUND 13
   1.1   Roadmap to This Report, 13
   1.2   Presidential Speech of January 2014 and PPD-28, 13
   1.3   Context for This Report, 15
   1.4   Legal and Policy Setting, 16
        1.4.1   The U.S. Constitution and the Legal and Regulatory Framework, 16
        1.4.2   Policy and Practical Controls, 22
        1.4.3   Legal Authorities for Collection and Use of Information, 22

2 BASIC CONCEPTS 26
   2.1   A Conceptual Model of the Signals Intelligence Process, 27
        2.1.1   Collection, 28
        2.1.2   Analysis, 31
        2.1.3   Dissemination, 32
   2.2   Bulk and Targeted Collection, 32
   2.3   Definitions of Critical Terms, 34

3 USE CASES AND USE CASE CATEGORIES 40
   3.1   Contact Chaining, 42
        3.1.1   Use Case 1, 42
        3.1.2   How Metadata Are Used in Contact Chaining, 43

3.2　Finding Alternate Identifiers, 44
　　　　　3.2.1　Use Case 2, 44
　　　　　3.2.2　Use Case 3, 45
　　　　　3.2.3　Use Case 4, 46
　　　　　3.2.4　How Metadata Are Used in Finding Alternate Identifiers, 47
　　　3.3　Triage, 48
　　　　　3.3.1　Use Case 4—Extension of the Scenario, 49
　　　　　3.3.2　Use Case 5—The Immediate Response After a Terrorist Incident, 49
　　　　　3.3.3　How Metadata Are Used in Triage, 49
　　　3.4　Conclusion, 49

4　BULK COLLECTION　　　　　　　　　　　　　　　　　　　51
　　　4.1　Uses of Bulk Collection, 51
　　　　　4.1.1　Information about the Past, 51
　　　　　4.1.2　Tactical Intelligence, 52
　　　　　4.1.3　Strategic Intelligence, 52
　　　　　4.1.4　Reference Data, 53
　　　　　4.1.5　Increasing the Likelihood That Needed Information Is Available, 54
　　　4.2　Alternatives to Bulk Collection, 54
　　　4.3　Conclusion, 56

5　CONTROLLING USAGE OF COLLECTED DATA　　　　　　59
　　　5.1　Why It Is Important to Control Usage, 59
　　　5.2　Controlling Usage, 60
　　　5.3　Manual Controls, 64
　　　5.4　Automatic Controls, 65
　　　　　5.4.1　Isolation, 66
　　　　　5.4.2　Restricting Queries Automatically, 75
　　　　　5.4.3　Audit/Oversight Automation, 75
　　　5.5　Conclusion, 76

6　LOOKING TO THE FUTURE　　　　　　　　　　　　　　　78
　　　6.1　The Future of Signals Intelligence, 78
　　　　　6.1.1　More Data, Data Types, and Sensors; More Computing and Storage, 79
　　　　　6.1.2　Business Records, 79
　　　　　6.1.3　Encryption, 80
　　　　　6.1.4　Services That Evade Surveillance, 81
　　　　　6.1.5　SIGINT Must Adapt, 81
　　　6.2　Evolution of Privacy Protections, 81

		6.3	Research and Development, 82
			6.3.1	Technologies for Isolation, 83
			6.3.2	Other Technologies for Protecting Data Privacy, 83
			6.3.3	Approving Queries and Their Results Automatically, 84
			6.3.4	Audit/Oversight Automation, 86
			6.3.5	Formal Expression of Laws and Regulations, 87
			6.3.6	Policy Research, 88
			6.3.7	Measuring Effectiveness of Intelligence Techniques and the Value of Data, 88
	6.4	Engagement with the Research Community, 89
	6.5	Conclusion, 90

APPENDIXES

A Observations about the Charge to the Committee   95
B Acronyms   97
C Biographical Information for Committee Members, Consultants, and Staff   99

Summary

This report of the Committee on Responding to Section 5(d) of Presidential Policy Directive 28: The Feasibility of Software to Provide Alternatives to Bulk Signals Intelligence Collection responds to a request to the National Academies from the Office of the Director of National Intelligence (ODNI). That request, in turn, was occasioned by Presidential Policy Directive 28 (PPD-28) Section 5(d), which had asked the Director of National Intelligence for "a report assessing the feasibility of creating software that would allow the Intelligence Community (IC) more easily to conduct targeted information acquisition rather than bulk collection [of signals intelligence]."[1] This study is among several of the administration's responses to heightened public concern about U.S. intelligence agency surveillance programs that followed Edward Snowden's disclosure of numerous internal National Security Agency (NSA) documents beginning in mid-2013. These responses include other activities called for in PPD-28 as well as in a study of big data and privacy by the President's Council of Advisors on Science and Technology that is largely focused on civilian applications.[2]

---

[1] The White House, Presidential Policy Directive/PPD-28, "Signals Intelligence Activities," Office of the Press Secretary, January 17, 2014, http://www.whitehouse.gov/sites/default/files/docs/2014sigint_mem_ppd_rel.pdf.

[2] President's Council of Advisors on Science and Technology, *Big Data and Privacy: A Technological Perspective*, Executive Office of the President, May 2014, http://www.whitehouse.gov/sites/default/files/microsites/ostp/PCAST/pcast_big_data_and_privacy_-_may_2014.pdf.

CONTEXT AND DEFINITIONS

PPD-28 defines bulk collection as "the authorized collection of large quantities of signals intelligence (SIGINT) data which, due to technical or operational considerations, is acquired without the use of discriminants (e.g., specific identifiers, selection terms, etc.)"[3] and implies that collection is targeted if it is not bulk. But PPD-28 defines "discriminant" only by example, so it does not provide a precise definition of either bulk or targeted collection. Nor are these terms defined precisely elsewhere in law or policy. Moreover, the PPD-28 description of bulk collection is problematic because it says that (1) with a broad discriminant, such as "Syria," collection is targeted, even though it captures a large volume of information and covers vast numbers of people who are not of intelligence value; and (2) if the signal itself contains only the traffic of a single individual, collection is bulk if there is no discriminant. Both of these results are inconsistent with the plain meaning of the words bulk and targeted.

Based in part on briefings from the IC, the committee adopted a definition better suited to understanding the trade-off between civil liberties and effective intelligence: *If a significant portion of the data collected is not associated with current targets,[4] it is bulk collection; otherwise, it is targeted.* There is no precise definition of bulk collection, but rather a continuum, with no bright line separating bulk from targeted. The committee acknowledges that use of the word "significant" makes its definition imprecise as well. The IC prefers targeted collection because it narrows its attention as much as possible during collection to use its limited resources efficiently, to comply with rules about what is allowed, and to limit intrusions on privacy.

This report, like PPD-28, focuses on a subset of SIGINT, a broad subset termed "communications or information about communications."[5] This includes electronic communications between people and those between people and services such as Internet search providers, message services, and banks. It also includes "business records" about communications. Intercepting these signals is of concern because it may intrude on the privacy and civil liberties of the communicators. However, this is only one ingredient among many that are used to meet the country's foreign intelligence needs. Understanding the nature of groups, individuals, organizations, or events that may threaten national security and predicting their behavior requires complex analysis that pieces together many facts from many sources. Studying this whole system was far beyond the scope of this study.

---

[3] Presidential Policy Directive/PPD-28, footnote 5.
[4] The term "target" and other key terms used in this report are defined in Section 2.3.
[5] Presidential Policy Directive/PPD-28, footnote 3.

The committee paid particular attention to collection of "information about communications," or metadata,[6] a focus of the briefings provided by the IC. NSA has been collecting metadata in bulk for domestic telephone calls since 2006; it has done so under the authority of Section 215 of the Foreign Intelligence Surveillance Act (FISA), enacted as part of the USA Patriot Act in 2001. This study applies not only to this practice but also to a broader set of activities, including the collection of metadata and contents of foreign telephone calls, emails, and other communications. This report addresses the question of alternatives to bulk collection, without regard to the specific authorities and restrictions that control the various types of bulk collection.[7]

This study, while focused on a technical question and on technological responses, inevitably encounters policy and privacy concerns; policy is bound to be affected by what is technically possible or impossible. Indeed, PPD-28 is itself a policy directive formed partly in response to privacy issues amplified by the Snowden disclosures. The committee did not study these policy questions and tried to avoid making judgments about them.[8] The committee tried to answer the technical question in general, rather than only in the context of current policy, because technology and policy can change rapidly.[9]

The next section provides a brief description of the SIGINT collection model used by the committee.

---

[6] In the case of telephone communications, "metadata" include the calling and called telephone numbers, the time and duration of a call, but not its content. For email, metadata have been interpreted to exclude the subject line. Other types of communications have different metadata elements.

[7] For example, FISA and Foreign Intelligence Surveillance Court (FISC) orders restrict bulk collection of domestic telephony records to querying targets with reasonable and articulable suspicion (RAS) that they belong to a foreign terrorist organization. For another example, PPD-28 restricts collection to six specific purposes.

[8] For recent reports that deal with policy associated with signals collection, see two reports from the Privacy and Civil Liberties Oversight Board: *Report on the Telephone Records Program Conducted under Section 215 of the USA Patriot Act and on the Operations of the Foreign Intelligence Surveillance Court*, January 23, 2014, http://www.pclob.gov/library/215-Report_on_the_Telephone_Records_Program.pdf, and *Report on the Surveillance Program Operated Pursuant to Section 702 of the Foreign Intelligence Surveillance Act*, July 2, 2014, http://www.pclob.gov/library/702-Report.pdf. See also President's Review Group on Intelligence and Communications Technologies, *Liberty and Security in a Changing World*, December 12, 2013, http://www.whitehouse.gov/sites/default/files/docs/2013-12-12_rg_final_report.pdf.

[9] Indeed, as this study was under way, the President announced he would seek legislation to end bulk collection of domestic telephony metadata (The White House, "The Administration's Proposal for Ending the Section 215 Bulk Telephony Metadata Program," Fact Sheet, March 27, 2014, Office of the Press Secretary, Washington, D.C.), and legislation was proposed.

A CONCEPTUAL MODEL OF SIGNALS INTELLIGENCE

In response to intelligence requirements determined by policy makers, NSA takes in signals,[10] extracts data about events, filters data according to one or more discriminants, stores the resulting data, analyzes it by querying the store, and disseminates the derived intelligence to other analysts and policy makers (Figure S.1). The first three steps are what the committee calls *collection*. The "extract" process decodes communications protocols to extract items for further inspection. A discriminant may be chosen to limit the collection to a set of targets determined at the time of collection; this is targeted collection. If a discriminant is chosen to collect a significant quantity of data not relevant to any current target, the collection is bulk. In either case, analysts query the data stored from multiple SIGINT collections and combine them with data from many other sources in order to formulate and disseminate intelligence useful to others. Privacy protections of different sorts are applied at various points throughout the process. These include choices about where to extract signals and what discriminants to use, minimization procedures used to protect information about U.S. persons, and controls on how collected information can be used.

Much of the data in the signal inevitably will *not* be of interest. This is because modern communication technology aggregates traffic between many sources and destinations onto a single channel—such as the fiber carrying Internet Protocol packets between two routers. With rare exceptions, there is no longer a single physical point, like the central office connection of a landline telephone, at which to observe exactly the items of interest. Thus, this definition of collection says that data is deemed collected only when it is stored for more than a few hours, not when it is extracted.

The distinction between bulk and targeted collection is not precise. When collection is very broad and it is expected that most of the information stored is not relevant to current targets, it is bulk. In contrast, if collection is about a person of interest, it is clearly targeted. There are, however, many cases in between. Throughout the intelligence process, agencies narrow their attention as much as possible, both to comply with rules about what is allowed and to use their limited resources efficiently. Narrowing applies to choosing signals from which to extract data, filtering the extracted data, querying collected data, and disseminating the results. For example, for domestic telephony metadata collected in bulk under the authority of FISA Section 215, a query is allowed only when

---

[10] The sources of the signals are a separate topic that the committee did not consider, although some examples are given later in the report.

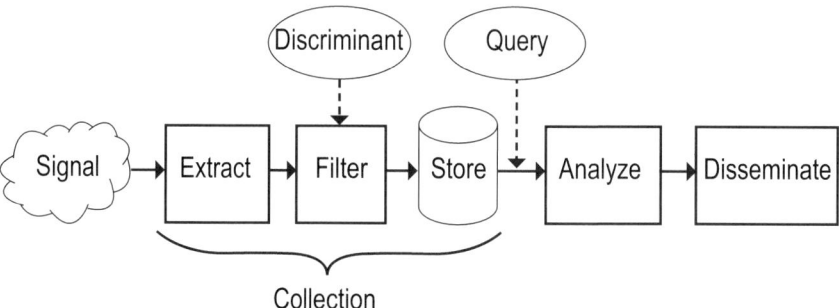

FIGURE S.1 A conceptual model for the signals intelligence process.

there is a reasonable and articulable suspicion that the target is associated with a foreign terrorist organization. Often, queries on bulk collections are sufficiently constrained that very little of the collected data is ever examined. Additional rules usually require collected data to be destroyed after a certain time.

CATEGORIES OF USE CASES

Use cases demonstrate how the results of intelligence analysis are used and make the process of intelligence more concrete for outsiders. Use cases that cover the full range of intelligence practice can provide confidence that the consequences of restricting bulk collection are understood and guide a search for alternatives. Although the committee was given unclassified use cases in three categories, it was told that this was not a complete set, so its search for collection alternatives was limited. The use case categories it was given, all of which concern communications between people who are designated by *identifiers* such as telephone numbers or email addresses, were the following:

- *Contact chaining*, which traces the network of people associated with a target by following links of the form "A communicated with B" starting at the target and traversing chains of one or more links.
- *Alternate identifier* techniques that seek to keep current the set of identifiers that a target person is known to be using, when the target is changing identifiers to avoid being tracked.
- *Triage* starts with a list of identifiers of interest and categorizes the urgency of the threat to national security from the party associated with each one.

A broader set of use cases, such as ones involving collection of communications content, detecting suspicious foreign communications patterns and suspicious queries to Internet search engines, might point to other possibilities for alternatives to bulk collection.

BULK COLLECTION AND INFORMATION ABOUT PAST EVENTS

A common aspect of the categories of use cases above is that they rely in part on information from the past to link or connect identifiers. If past events become interesting in the present—because of new circumstances such as identifying a new target, a nonnuclear nation that is now pursuing the development of nuclear weapons, an individual who is found to be a terrorist, or new intelligence-gathering priorities—then historical events and the data they provide will be available for analysis only if they were previously collected. If it is possible to do targeted collection of similar events in the future, and if they happen soon enough, then the past events might not be needed. If the past events are unique or if delay in obtaining results is unacceptable (because of an imminent threat or perhaps because of press coverage or public demand), then the intelligence will not be as complete. So restricting bulk collection will make intelligence less effective, and technology cannot do anything about this; whether the gain in privacy is worth the loss of information is a policy question that the committee does not address.

CONTROLLING USAGE

Controls on usage can help reduce the conflicts between collection and privacy. There are other entities that collect highly sensitive data and use it for purposes that the people who provide it might not like, such as companies that provide cloud services such as email and social media and "data brokers" that collect and correlate data from a wide variety of public and proprietary sources and sell it to help with decisions about extending credit or for marketing purposes. It is worth comparing how society controls these activities with how it controls the IC. The accepted control paradigm is "notice and consent," the terms of service that almost no one reads. Although today people are more tolerant of private data collection than of government data collection, this may change as the collection of private data grows. The 2014 report on privacy and big data from the President's Council of Advisors on Science and Technology proposes instead that people should have control over how their data are used.[11] Controls on use thus offer an alternative to controls on collection as a way of protecting privacy.

There are two ways to control usage: manually and automatically. NSA already has both automated and strong manual controls in place. Despite rigorous auditing and oversight processes, however, it is hard to convince outside parties of their strength, because necessary secrecy prevents the public from observing the controls in action, and because popular descriptions of the controls are imprecise and sometimes wrong. Technical means can isolate collected data and automatically restrict queries that analysts make, and the way these means work can be public without revealing sensitive sources and methods. Then people outside the IC concerned about privacy and civil liberties would have new ways to verify that the IC has adequate procedures and follows them. Enhanced automated controls also offer the promise of reduced burdens on analysts because they can be more efficient than manual controls. Some manual controls would still be necessary to ensure that the automatic controls are actually imposed and that they are configured according to the rules, and to decide cases that are too complex to be automated.

Automated controls and audits require expressing, in software, the rules embodied in laws, policies, regulations, and directives that constrain how intelligence is collected, analyzed, and disseminated. The current rules form a complex network that has grown with changes in technology and in the national security environment. They contain conflicting definitions and inconsistencies. Deriving from the legislative and administrative expressions of the rules, an expression in a concise, consistent, machine-processable form would not only simplify automation software but also make the rules more understandable to the public.

The next section outlines the key technical elements required to control and automate usage.

TECHNICAL ELEMENTS OF AUTOMATED CONTROLS

An automated system for controlling usage of bulk data with high assurance has three parts: isolating bulk data so that it can be accessed only in specific ways, restricting the queries that can be made against it, and auditing the queries that have been done. In each of these areas, there are opportunities for automated control; some of them are already

---

[11] President's Council of Advisors on Science and Technology (PCAST), *Big Data and Privacy: A Technological Perspective,* Executive Office of the President, May 2014, http://www.whitehouse.gov/sites/default/files/microsites/ostp/PCAST/pcast_big_data_and_privacy_-_may_2014.pdf.

deployed in the IC or in private companies, some have been demonstrated in research laboratories, and some are promising research directions.

Isolating bulk data is one technical method for controlling usage. Figure S.2 shows the elements of this method. Bulk data are cut off from the outside world by an isolation boundary. The only way to cross this boundary is to submit a query to the guard, which enforces the policy that controls what queries and results are allowed. The guard logs all queries and results for later auditing, and the audit log itself is isolated to protect it from tampering. The isolated domain is hosted by a mechanism that guarantees the isolation. The guard, the isolation boundary, and bulk data processing are the critical parts of this system. The simpler and clearer their tasks are, and the shorter and clearer the software programs that implement them, the more likely they are to be trustworthy.

Restricting queries automatically in the guard is another aspect of controlling usage automatically. The goal is to do this well enough that software can decide which queries are allowed by the policy, or at least drastically reduce the number of queries that require manual, human approval. This is certainly feasible for limited classes of queries such as, "Find all the phone numbers that have connected in the last month to this list of numbers belonging to a known target." Indeed, NSA already has pre-approved queries.

Auditing usage of bulk data is essential to enforce privacy protections. Isolation provides confidence that every query is permanently logged and that the log cannot be altered. Then the log must be reviewed for compliance with the rules. Doing this manually is feasible and is, indeed, NSA's current practice. Although it is thorough, it is expensive and not

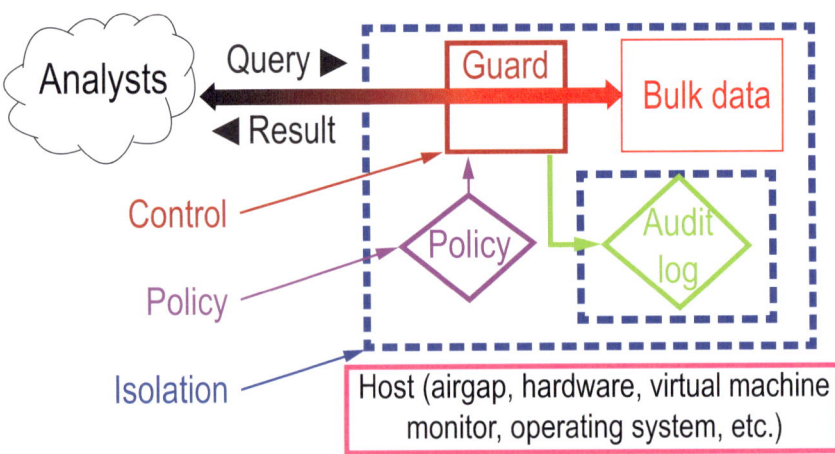

FIGURE S.2 Controlling usage by isolating bulk data.

transparent—outsiders must rely on the agency's assurance that it is being done properly because the queries are usually highly classified. Automation of auditing, a direction NSA is pursuing, could not only streamline audits but also provide assurance to outside inspectors, who can examine the auditing technology. Automation of auditing is an area that has been neglected by government, industry, and academia.

Automated controls and auditing of SIGINT data held and accessed securely may allow sufficiently thorough unclassified inspection of the privacy-protecting mechanisms of the SIGINT process to allay privacy and civil liberty concerns. The inspection would focus on the automation software and the usage rules it enforces rather than on the data, which must remain classified.

CONCLUSIONS

Although no software can fully replace bulk with targeted information collection, software can be developed to more effectively target collection and to control the usage of collected data.

Conclusion 1. There is no software technique that will fully substitute for bulk collection where it is relied on to answer queries about the past after new targets become known.

A key value of bulk collection is its record of past SIGINT that may be relevant to subsequent investigations. If past events become interesting in the present, because intelligence-gathering priorities change to include detection of new kinds of threats or because of new events such as the discovery that an individual is a terrorist, historical events and the context they provide will be available for analysis only if they were previously collected.

The committee was not asked to and did not consider whether the loss of effectiveness from reducing bulk collection would be too great or whether the potential gain in privacy from adopting an alternative is worth the potential loss of intelligence information. Nor was it able to identify broad categories of use where substitution of alternatives might be possible or to detect broadly useful metrics that would inform such decisions. ODNI may wish to study these questions further.

Other groups, such as the President's Review Group on Intelligence and Communications Technologies and the Privacy and Civil Liberties Oversight Board (in its Section 215 report), have said that bulk collection of telephone metadata is not justified.[12] These were policy and legal judgments that are not in conflict with the committee's conclusion that there is

---

[12] See footnote 8.

no software technique that will fully substitute for bulk collection; there is no technological magic.

Conclusion 1.1. Other sources of information might provide a partial substitute for bulk collection in some circumstances.

Data retained from targeted SIGINT collection is a partial substitute if the needed information was in fact collected. Bulk data held by other parties, such as communications service providers, might substitute to some extent, but this relies on those parties retaining the information until it is needed, as well as the ability of intelligence agencies to collect or access it in an efficient and timely fashion. Other intelligence sources and methods might also be able to supply some of the lost information, but the committee was not charged to and did not investigate the full range of such alternatives. Note that these alternatives may introduce their own privacy and civil liberties concerns.

Conclusion 1.2. New approaches to targeting might improve the relevance of the collected information to future use and would rely on capabilities such as creating and using profiles of potentially relevant targets, possibly by using other sources of information.

Because bulk collection cannot for practical reasons be truly comprehensive, it is itself inherently selective and unable to capture all relevant history.[13] It may be possible to improve targeted collection to the point where it provides a viable substitute for bulk collection in at least some cases, using profiles of potential targets that are compiled from a wide range of information. This might reduce collection against persons who are not targets, but it might also introduce new privacy and civil liberties concerns about how such profiles are developed and used.

Rapidly updating discriminants of ongoing collections to include new targets as they are discovered will collect data that would otherwise be lost. If targeted collection can be done quickly and well enough, bulk information about past events may not be needed. Targeted collection cannot be a substitute if the past events were unique or if the delay incurred to collect new information would be unacceptable.

Conclusion 2. Automatic controls on the usage of data collected in bulk can help to enforce privacy protections.

---

[13] The FISA Section 215 program collects "only a small percentage of the total telephony metadata held by service providers" (President's Review Group on Intelligence and Communications Technologies, *Liberty and Security in a Changing World,* 2013, p. 97).

Automation of usage controls may simultaneously allow a more nuanced set of usage rules, facilitate compliance auditing, and reduce the burden of controls on analysts. Similarly, there are opportunities to automate the various audit mechanisms to verify that rules are followed. Such capabilities could be enhanced as the information technology systems for collection and analysis are refreshed and modernized. These techniques may permit more of the use controls and audit mechanisms to be explained clearly to the public. It may be possible to express a large fraction of the rules required by law and policy in a machine-processable form and thus apply them rapidly and consistently during collection, analysis, and dissemination.

> **Conclusion 2.1.** It will be easier to automate controls if the rules governing collection and use are technology-neutral (i.e., not tied to specific, rapidly changing information and communications technologies or historical artifacts of particular technologies) and if they are based on a consistent set of definitions.[14]

> **Conclusion 2.2.** Automated controls can provide new opportunities to make the controls more transparent by giving the public and oversight bodies the opportunity to inspect the software artifacts that describe and implement the controls. Increased transparency can give people outside the IC more confidence that the controls are appropriate, although the need for secrecy about some of the details makes complete confidence unlikely.

> **Conclusion 3.** Research and development can help in developing software intended to (1) enhance the effectiveness of targeted collection and (2) improve automated usage controls.[15]

> **Conclusion 3.1** The use of targeted collection can be improved by enriching and streamlining methods for determining and deploying new targets rapidly, using automated processing and/or streamlined approval procedures.[16]

---

[14] This conclusion is consistent with Recommendation 2 in PCAST, *Big Data and Privacy: A Technological Perspective*, 2014.

[15] See also Ibid., Recommendation 3.

[16] Examples of manual procedures for target approval are in National Security Agency, *NSA's Civil Liberties and Privacy Protections for Targeted SIGINT Activities Under Executive Order 12333*, NSA Director of Civil Liberties and Privacy Office Report, October 7, 2014, https://www.nsa.gov/civil_liberties/_files/nsa_clpo_report_targeted_EO12333.pdf.

Analytics, such as "big data analytics," may help narrow collection, even if they are not sufficiently precise to identify individual targets. If the government is constrained by privacy concerns to collect less data, it may nevertheless be able to use the power of large private-sector databases, analytics, and machine learning to shape the constraints to collect only data predicted to have high value. New uses by the government of private-sector databases would also raise new privacy and civil liberties questions.

Advanced targeting methods may require a great deal of computing, so that filters should be cascaded to first apply cheap tests, followed by more expensive filters only if earlier filters warrant. For example, if metadata indicate a civilian telephone call to a military unit under surveillance, speech recognition and subsequent semantic analysis might be applied to the voice signal, resulting in an ultimate collection decision. Richer targeting may require enhancing the ability of collection hardware and software to apply complex discriminants to real-time signals feeds.

Conclusion 3.2. More powerful automation could improve the precision, robustness, efficiency, and transparency of the controls, while also reducing the burden of controls on analysts.

Some of the necessary technologies exist today, although they may need further development for use in intelligence applications; others will require research and development work. This approach and others for privacy protection of data held by the private sector can be exploited by the IC.[17] Research could also advance the ability to systematically encode laws, regulations, and policies in a machine-processable form that would directly configure the rule automation.

It does not necessarily follow from Conclusion 1 that current bulk collection must continue. What it does mean is that curtailing bulk collection would deprive analysts of some information. Reduction in bulk collection may be partially mitigated by improvements in targeting, a direction for future research outlined above. If the IC continues to collect SIGINT in bulk, the technology described in this report can reduce risk and improve oversight and transparency and, thus, perhaps mitigate public concerns about it.

---

[17] PCAST, *Big Data and Privacy: A Technological Perspective*, 2014. Recommendation 1, Sections 4 and 4.5.2.

1

Introduction and Background

1.1 ROADMAP TO THIS REPORT

This chapter provides the basic context in which this report is being written. Chapter 2 of this report introduces some basic concepts of signals intelligence (SIGINT) and provides some key definitions. Chapter 3 presents use cases—scenarios in which bulk collection may make contributions to intelligence investigations. Chapter 4 presents the committee's technical conclusions about the use of bulk collection. Chapter 5 describes ways of protecting information gathered through SIGINT processes. Chapter 6 looks to the future. Appendix A makes some observations about how the committee addressed its charge.

1.2 PRESIDENTIAL SPEECH OF JANUARY 2014 AND PPD-28

In January 2014, President Obama addressed the nation and the broader global community to explain U.S. policy regarding the collection of foreign intelligence.[1] In this speech, he explicitly acknowledged that U.S. government collection and storage of bulk data "creates a potential for abuse," but he explained that signals intelligence data were collected only for "legitimate national security purposes" and that the government had no interest in using any collected data to target minorities or

---

[1] The White House, "Remarks by the President on Review of Signals Intelligence," Office of the Press Secretary, January 17, 2014, http://www.whitehouse.gov/the-press-office/2014/01/17/remarks-president-review-signals-intelligence.

suppress any political activity. He clarified that the use of any bulk collection of SIGINT was even more limited, explicitly stating that it could be used only for six specific security requirements: "counterintelligence; counterterrorism; counterproliferation; cybersecurity; force protection for our troops and our allies; and combating transnational crime, including sanctions evasion."

While defending the nature of American collection and use of bulk data to support national security, the President also acknowledged how many in America and around the world might still be concerned. He declared an interest in exploring how the United States can preserve current intelligence capabilities but with less government collection and storage of bulk data. He conceded that it would not be easy to "match the capabilities and fill the gaps that the [metadata[2] collection program] was designed to address," but he is committed to exploring several options that might enhance protections of privacy, including decreasing the number of "hops" in a contact network search to two from three, having the Foreign Intelligence Surveillance Court (FISC) review reasonable and articulable suspicion (RAS) selectors and identifying a means to have the storage of the bulk metadata occur outside the federal government.

Shortly after the President's speech, the White House released Presidential Policy Directive 28 (PPD-28),[3] the topic of which was U.S. policy on SIGINT. PPD-28 both laid out the principles that govern how the U.S. collects SIGINT and strengthened executive branch oversight of SIGINT activities. PPD-28 seeks to ensure that U.S. policy takes into account security requirements, alliances, trade and investment relationships (including the concerns of U.S. companies), and the U.S. commitment to privacy and basic rights and liberties. The document also promised review of U.S. decisions about intelligence priorities and sensitive targets by the President's senior national security team on an annual basis.

Of most importance to this report, PPD-28 requested the Office of the Director of National Intelligence (ODNI) to "assess the feasibility of creating software that would allow the Intelligence Community more easily to conduct targeted information acquisition rather than bulk collection." In turn, ODNI asked the National Academies to study and report on this question. The Committee on Responding to Section 5(d) of Presidential Policy Directive 28: The Feasibility of Software to Provide Alternatives to Bulk Signals Intelligence Collection was formed in response.

---

[2] The term metadata is defined in Section 2.3. Loosely, for telephone calls it includes calling and called number, and time and duration of call, but not any content of the call.

[3] See The White House, Presidential Policy Directive/PPD-28, "Signals Intelligence Activities," Office of the Press Secretary, January 17, 2014, http://www.whitehouse.gov/the-press-office/2014/01/17/presidential-policy-directive-signals-intelligence-activities.

1.3 CONTEXT FOR THIS REPORT

The broader context for the committee's report includes the many security threats the United States faces, issues of international relations and global competitiveness, balanced against privacy and civil-liberties concerns. The committee believes that both the public and national security officials recognize the need for surveillance to anticipate, disrupt, and respond to national security threats (such as terrorism). Indeed, recent events make clear that national security threats will continue to be dynamic and more unpredictable than those during the Cold War, so that effective intelligence capabilities will remain essential.

At the same time, disclosures by Edward Snowden about the extent and nature of U.S. intelligence collection have raised concerns about the appropriate balance between the surveillance needed to achieve national security and respect for individual privacy. The revelations have complicated U.S. relations with other nations. (This is true even as some of these same nations have benefitted—and continue to benefit—from U.S. intelligence collection.) A number of foreign nations have threatened to avoid Internet-delivered products and services offered by U.S. information technology vendors because of insecurities alleged in these disclosures.[4] The magnitude of the financial impact is unclear at this point;[5] what is clear is that increased attention to U.S. intelligence collection has made the international marketplace a more challenging environment for U.S. companies.

The Snowden disclosures have also generated a range of concerns about privacy and civil liberties. In the United States, tension over potential government infringement of personal liberties goes back to the founding of the republic. In the recent past, domestic legislation and case law have worked to create a balanced approach to surveillance of telephone communications. But new technologies—and how government authorities use such technologies—have always posed challenges for existing law and practice. Technological advancements can undo a previously agreed-upon consensus. In the controversy that gives rise to this report, domestic public concerns about privacy and civil liberties have often been expressed as concerns about the "[U.S.] government spying on innocent Americans." Abroad, a Pew poll in July 2014 indicated that in most coun-

---

[4] Whether this reason is in some sense "sincere," or a cover for protectionism is unclear. But it may not matter. Whether perception or reality, U.S. leadership has concluded that the upset created is sufficient to require a response.

[5] For example, a *New York Times* article of March 2014 reports on estimates of losses to U.S. technology companies ranging from $35 billion to $180 billion by 2016. See Claire Cain Miller, "Revelations of N.S.A. Spying Cost U.S. Tech Companies," *The New York Times*, March 21, 2014, http://www.nytimes.com/2014/03/22/business /fallout-from-snowden-hurting-bottom-line-of-tech-companies.html?_r=0.

tries surveyed, the majority of their publics opposed U.S. surveillance of their citizens or their leaders.[6] Foreign leadership, including that of traditionally close U.S. allies, expressed significant public anger in the immediate aftermath of the Snowden disclosures.[7]

Recent disclosures have amplified two underlying trends in the United States: distrust of government and concerns about privacy, especially privacy of data and communications. These trends put pressure on SIGINT techniques and practices, which in turn may affect the quality of intelligence they provide. But another attack of the magnitude of the 9/11 attacks would quickly raise expectations for the capabilities of U.S. intelligence generally and surveillance in particular. Actions by Congress and the executive branch in the wake of the 9/11 attacks rapidly resulted in substantive changes that are now being questioned after more than a decade of relative domestic security. As this cycle may easily repeat, it is all the more imperative to examine now the value of bulk collection and potential alternatives while there is time to reflect thoughtfully on the issues such collection poses.

1.4 LEGAL AND POLICY SETTING

1.4.1 The U.S. Constitution and the Legal and Regulatory Framework

As the committee proceeded with its work, it became clear that public confidence in the management of intelligence programs is essential and might be enhanced through even greater use of automation in managing oversight structures. To provide the context for the committee's subsequent discussion, particularly on ways to automate oversight strategies, this section provides a brief overview of the constitutional and legal framework that currently governs intelligence surveillance activities. Mindful that many long-standing legal interpretations are now under review by Congress and have also been challenged in lower federal courts, and that there has been no final disposition of these questions by either Congress or the Supreme Court, the committee does not discuss the current legal debate in depth.

Among the nations of the world (including the Western democracies), the United States is the most open in the regulation of its intelligence activities. The United States regulates its intelligence activities

---

[6] Pew Research Center, *Global Opposition to U.S. Surveillance and Drones, but Limited Harm to America's Image*, Washington, D.C., July 14, 2014, http://www.pewglobal.org/2014/07/14/global-opposition-to-u-s-surveillance-and-drones-but-limited-harm-to-americas-image/.

[7] Josh Levs and Catherine E. Shoichet, Europe furious, 'shocked' by report of U.S. spying, *CNN*, July 1, 2013, http://www.cnn.com/2013/06/30/world/europe/eu-nsa/.

according to a legal framework established by the U.S. Constitution. This overarching constitutional structure is premised on the commitment that all governmental activities, even those national security activities most important to the nation's existence, must be subject to the rule of law. This framework is further explicated by a hierarchy of public statutes and internal executive branch regulations, which include public executive orders and subordinate classified instructions and directives.

All three branches of government have a constitutional role to play in intelligence programs. The executive branch is responsible for executing intelligence programs; congressional committees have responsibility for the initial authorization, funding, and oversight of programs; and the federal courts provide legal review in the course of litigation and also, in a limited number of cases, prior authorization through the specially created FISC.[8] Authoritative descriptions of the legal constraints imposed on U.S. intelligence functions are available elsewhere,[9] but a brief outline of this legal framework will help illustrate how U.S. intelligence programs function and facilitate the discussion that follows about various programs conducted by the National Security Agency (NSA).

Article II of the U.S. Constitution assigns three functional roles to the President: Commander in Chief, responsibility for the conduct of foreign affairs, and at home, execution of the laws. The responsibilities as Commander in Chief and for foreign affairs carry with them an inherent constitutional power to gather intelligence. Like all such constitutionally granted powers, limits contained in the Bill of Rights amendments to the Constitution apply. With regard to intelligence, particularly the SIGINT for which NSA is responsible, two are particularly relevant: the Fourth Amendment, which protects individuals against unreasonable searches and seizures; and the First Amendment, which protects freedom of speech and assembly, as well as freedom of the press.

The U.S. Supreme Court has interpreted the Fourth Amendment's protections against a standard of reasonableness so that an individual's privacy interest must be weighed against the legitimate interests of the government for national security and public safety. In addition, the Amendment has differential applications depending on the purpose of the surveillance, where it occurs (e.g., inside or outside the United States), and the subject of the surveillance (e.g., a non-U.S. person or a U.S. person outside the United States). Finally, the committee notes that privacy

---

[8] Foreign Intelligence Surveillance Act of 1978, 50 U.S.C., Ch. 36 (1978), as amended.

[9] Robert S. Litt, "Privacy, Technology and National Security: An Overview of Intelligence Collection," speech, Washington, D.C., July 18, 2013, http://www.dni.gov/index.php/newsroom/speeches-and-interviews/174-speeches-interviews-2009. See also Dycus, Banks, Raven-Hansen, Vladeck, "National Security Law" (5th edition) and "Counterterrorism Law" (2nd edition), Wolters Kluwer, 2014-2015 supplement.

interests may be limited insofar as information is shared voluntarily with others.[10]

NSA was originally created by presidential memorandum under the statutory authority of the Department of Defense to create combat support agencies. By contrast, most other agencies that implement the President's intelligence needs have been created directly by Congress pursuant to their own explicit "organic" statutes. All, however, may only act insofar as they are authorized to do so, whether by statute, regulation, or executive order. In some cases, they are specifically prohibited from action. For example, by statute the Central Intelligence Agency may not conduct domestic law enforcement. NSA, as a part of the Department of Defense, has no separate authorizing statute, but the same prohibition applies to it by regulation.

As has already been noted, law sometimes lags behind changes in technology. One example is that wiretaps were not considered subject to Fourth Amendment protections until 1967 when the Supreme Court concluded that a right of individual privacy existed to protect against warrantless searches.[11] In subsequent years, an increasing number of laws have been passed at both the national and state levels to regulate the ways in which the government, including its intelligence components, may make use of evolving telecommunications and computer technologies.

For this report, focused on NSA programs, among the most significant is the Foreign Intelligence Surveillance Act (FISA) of 1978, later amended in 2001 and 2008 to add authorities. Although FISA establishes specific procedures to govern intelligence collection activities that involve U.S. citizens or territory,[12] NSA's institutional charter is found in Executive Order 12333 and is further defined by guidelines called U.S. Signals Intel-

---

[10] Significant for some metadata collection programs, in an opinion authored by Justice Harry Blackmun, the U.S. Supreme Court found in *Smith v. Maryland*, 442 U.S. 735, 1979, that no "legitimate expectation of privacy" existed if a third party, such as the phone company, already had access to information. Thus, because the phone company had retained the numbers of calls made, collecting them with a "pen register" was not a Fourth Amendment search requiring prior court authorization by warrant. Some question whether Smith remains "good law" today in light of the differing technology involved in modern metadata collection; this is the subject of current litigation.

[11] Compare *U.S. v. Olmstead*, 277 U.S. 438, 1928, holding no Constitutional protection for phone conversations with *Katz v. United States*, 389 U.S. 347, 1967, finding a right to privacy under the Fourth Amendment for the content of such communications.

[12] Foreign Intelligence Surveillance act of 1978 (http://www.gpo.gov/fdsys/pkg/STATUTE-92/pdf/STATUTE-92-Pg1783.pdf), 50 U.S.C. § 1881a, 1978, and brief description of its provisions (http://www.gpo.gov/fdsys/pkg/BILLS-110hr6304enr/pdf/BILLS-110hr6304enr.pdf).

ligence Directives (USSID), the most important of which for this report is USSID 18, which has been declassified in substantial part.[13]

The original enactment of FISA responded to significant contemporary political pressures, which resulted from abuses revealed in a series of congressional hearings in the 1970s, and demanded greater control of foreign intelligence collection by SIGINT methods when an activity occurs in the United States or involves U.S. persons. The level of statutory and regulatory control responds to political pressures that ebb and flow over time; as will be seen. The 9/11 attacks caused an adjustment in this balance to respond to foreign attacks in domestic space.

At its initial enactment, FISA was not without controversy. Although some argued that there was a critical need for the oversight that FISA provided through a specially created court, others argued (and continue to do so today) the long-standing view that foreign intelligence, as a core presidential function, could not be regulated constitutionally by congressional statute.[14] Nonetheless, passage of FISA, which introduced court approval of intelligence collection for the first time, was encouraged by a contemporaneous decision of the U.S. Supreme Court, intimating that much of such domestic national security collection might be subject to Fourth Amendment requirements for prior judicial approval through a warrant application process.[15] In response, FISA created a unique procedural approval process overseen by a new Article III court, the FISC, which was designed to authorize electronic intelligence surveillance in the

---

[13] National Security Agency, United States Signals Intelligence Directive USSID SP0018, (U) Legal Compliance and U.S. Persons Minimization Procedures, Issue Date January 25, 2011, approved for release on November 13, 2013, referred to as USSID 18, http://www.dni.gov/files/documents/1118/CLEANEDFinal USSID SP0018.pdf.

[14] A recently released May 6, 2004, Memorandum for the Attorney General authored by Professor Jack L. Goldsmith, then Assistant Attorney General, Department of Justice, Office of Legal Counsel, describes this view. See Jack L. Goldsmith, *Review of the Legality of the STELLAR WIND Program*, Office of the Assistant Attorney General, Washington, D.C., May 6, 2004, http://www.justice.gov/sites/default/files/pages/attachments/2014/09/19/may_6_2004_goldsmith_opinion.pdf.

[15] Although Title III of the Omnibus Crime Control and Safe Streets Act of 1968, 18 U.S.C. §§2510-2520, 1968, authorizes electronic surveillance for specifically limited crimes with a prior court order, a proviso at 18 U.S.C. §2511(3) protected the President's long-standing right to conduct surveillance for national security purposes. Nonetheless, Justice Lewis Powell's language in the majority decision of *U.S. v. United States District Court (Keith)*, 407 U.S. 297, 1972, had made clear that this exception would be narrowly construed in cases of domestic security. FISA responded to indications of the direction of Supreme Court decisions. In the Keith decision, it was argued that the defendants, U.S. citizens who had acted only domestically, constituted national security threats by bombing a government facility and so the warrant requirement of the Fourth Amendment did not apply. The Supreme Court rejected this contention, but left open the possibility that the executive branch might not be so limited if national security threats involved foreign powers.

United States by NSA and the Federal Bureau of Investigation upon application to, and approval by, the court. The FISC decisions have remained largely classified throughout much of the court's history. This proved controversial to some. They questioned the independence of a judicial body that operated largely out of the public eye to authorize intrusive surveillance that, unlike warrants in criminal matters, would likely never be publicly available, lacked any adversarial process, and limited the right of appeal to the government applicant alone. These questions remain and provide part of the backdrop to this report.

As originally enacted, FISA governed electronic surveillance for foreign intelligence or counterintelligence information when collection would occur within the United States. To collect such information, a showing must be made to the FISC establishing probable cause that the target is either a foreign power or an agent of a foreign power. Where the target is a U.S. person, a showing based solely on First Amendment activities is not sufficient. Collection is subject to *minimization* protections, procedures designed to limit "the acquisition and retention, and prohibit the dissemination, of nonpublicly available information concerning unconsenting United States persons," but in ways nonetheless consistent with the need for foreign intelligence.[16] As a practical matter, minimization involves removing the names of and references to U.S. persons with these exceptions: the information is necessary to assess the value of the foreign intelligence or the targeting of a U.S. person was approved by the FISC.

FISA was amended following the collection of domestic communications metadata that began in 2001. This was done initially at presidential direction outside normal FISA processes, a decision that proved controversial.[17] It was subsequently brought within the FISA process in 2006 through the "business records" provision of Section 215 of the USA Patriot Act.[18] This allowed the FISC to require production of documents and other tangible things determined relevant to national security investigations, much as other courts do in criminal and grand jury investigations. This provision has served as the authority under which the U.S. government has requested telecommunications providers to produce telephony metadata, when relevant to a national security investigation.[19] This provi-

---

[16] See Foreign Intelligence Surveillance Act of 1978, 50 U.S.C. §§1801(h)(1) and 1821(4)(A), 1978.

[17] See footnote 14.

[18] USA Patriot Act 2001, http://www.gpo.gov/fdsys/pkg/PLAW-107publ56/pdf/PLAW-107publ56.pdf.

[19] Standards of relevance vary according to context. What is relevant for a criminal investigation will differ from the far broader standard for civil discovery or a grand jury subpoena. The FISC has acceded to the government's argument that for national security investigations, "relevance" must be broadly construed. See Robert S. Litt, "Privacy, Technology and National Security," 2013, p. 6.

sion, approved in the course of several reviews by the FISC since 2006, was also reauthorized by Congress in 2009 and again in 2011. It should be noted that the interpretation of Section 215 permitting bulk collection of such business records, although provided to Congress and relevant committees, was not publicly acknowledged by the U.S. government until after the Snowden disclosures.[20]

A third provision was added when Section 702 was passed as part of the FISA Amendments Act of 2008 and reauthorized in 2012.[21] The Section 702 amendment brought all communications, whether by satellite, radio, wire, etc., acquired with the assistance of electronic communication service providers under FISC oversight and supervision, even though these communications were occurring overseas. Section 702 allows the targeting of non-U.S. persons who are reasonably believed to be outside the United States and expected to possess, receive, and/or communicate foreign intelligence information, consistent with the Fourth Amendment.

Although full communications content, not just metadata, can be collected under this authority, only non-U.S. persons may be targeted for approved foreign intelligence purposes. To ensure that these limitations are followed while preserving the flexibility and nimbleness needed for effective foreign intelligence collection, annual certifications by the U.S. Attorney General are presented to the FISC for approval, rather than specific prior judicial approval on a case-by-case basis.

The foregoing FISA provisions do not fully describe NSA's collection authority. To ensure that all collection was consistent with constitutional requirements, a broad operational "charter," Executive Order 12333, "United States Intelligence Activities," was promulgated in 1981 by the Reagan Administration; this has continued without significant change in collection authorities until the present. This executive order provides the basic authorities and principles under which all national security agencies must operate.[22] Importantly, at §2.8, "Consistency with Other Laws," it provides: "Nothing in this Order shall be construed to authorize any activity in violation of the Constitution or statutes of the United States." The provisions of Executive Order 12333 are further supported by detailed operating regulations applicable to each individual agency; in the case of NSA, Department of Defense Regulation 5240.1-R, its clas-

---

[20] David S. Kris, On the bulk collection of tangible things, *Journal of National Security Law and Policy* 7:209, 2014.

[21] FISA Amendments Act of 2008, http://www.gpo.gov/fdsys/pkg/BILLS-110hr6304enr/pdf/BILLS-110hr6304enr.pdf.

[22] Executive Order 12333, http://www.archives.gov/federal-register/codification/executive-order/12333.html. NSA's 13 specified responsibilities are defined at Executive Order No. 12333 §1.12(b), 3 C.F.R. 200, 1981, Intelligence Components Utilized by the Secretary of Defense.

sified annex, and USSID 18, approved by the Attorney General, provide the specific implementation guidance for all authorized activities.

USSID 18 offers an important window into the detailed operational authorities that govern NSA activities.[23] It begins by observing that all NSA activities must be consistent with the Constitution's provisions, as interpreted by the U.S. Supreme Court. Annex A to USSID 18 sets forth minimization procedures approved by the Attorney General that govern the handling of information under FISA authority that may relate to U.S. persons. The procedures limit the retention and dissemination of information about U.S. persons, whether or not the information is pertinent. Incidental collection of data about individuals who are not themselves subjects of interest is common to all forms of collection, and the concept of minimization is thus one of long standing in law enforcement activities.

1.4.2 Policy and Practical Controls

Responding to the legal framework described above, NSA has developed a system of internal compliance and oversight. All parts of the foreign intelligence collection system are involved: access, storage, analysis, and dissemination.

Both manual and automated controls are used to implement the legal search framework that governs foreign intelligence information. Controls and secure databases are used next to protect the subsequent storage of foreign intelligence information. Subsequent review of all actions is extensive. An automatically generated audit trail and internal and external human review are involved. Extensive training for all NSA employees also occurs.

An example of how policy and practical controls work together to protect privacy in the case of data gathered under Section 215 authority is provided in Box 1.1.

1.4.3 Legal Authorities for Collection and Use of Information

The legal authorities under which NSA operates are described in a public document entitled *NSA Missions, Authorities, Oversight and Partnerships*.[24] As noted above, these authorities include Executive Order 12333 and the Foreign Intelligence Surveillance Act of 1978, as amended. Executive Order 12333 is the foundational authority on which NSA relies to collect, retain, analyze, and disseminate foreign SIGINT information.

---

[23] See USSID 18.

[24] National Security Agency, *The National Security Agency: Missions, Authorities, Oversight and Partnerships*, August 9, 2013, https://www.nsa.gov/public_info/_files/speeches_testimonies/2013_08_09_the_nsa_story.pdf.

BOX 1.1
Privacy Protections for Phone Metadata Collected Under Section 215

Privacy protections for telephone metadata collected under Section 215 authority were described in a speech by Office of the Director of National Intelligence (ODNI) General Counsel Robert Litt on July 18, 2014.[a,b] He noted that before reports from queries are returned to analysts, the queries themselves must be approved to ensure compliance with legal and policy rules. These rules may stem from law (e.g., Section 215 restrictions on surveillance of U.S. persons) or from internal controls (e.g., that an analyst must be trained on the proper use of the returned data). All queries must meet a reasonable and articulable suspicion test. These rules seek to ensure that there can be no domestic "fishing expeditions" in which queries seek information about parties unrelated to an intelligence investigation.

Litt also reported on other measures that are applied to protect privacy of Section 215 telephone metadata:

- The information is stored in secure databases.
- The only intelligence purpose for which the information can be used is counterterrorism.
- Only a limited number of analysts may search these databases.
- A search is allowed only when there is already a reasonable and articulable suspicion that the telephone number is associated with a terrorist organization that has been identified by the FISC.
- The data may be used only to map a network of telephone numbers calling other telephone numbers.
- If an analyst finds a previously unknown (domestic) telephone number that warrants further investigation, that number may only be disseminated in a way that avoids identifying a person associated with the number. Further investigation may be done only by other lawful means, including other FISA provisions and law enforcement authority.
- The telephony metadata is destroyed after 5 years.
- Audit records are kept for all database queries, and a set of auditing and compliance-checking procedures applies, implemented by not only NSA but also ODNI and the Department of Justice.

In addition, only a limited number of NSA officials (22) are designated to make a determination that a telephone number satisfies the reasonable and articulable suspicion (RAS) criteria.[c]

---

[a] Robert S. Litt, "Privacy, Technology and National Security: An Overview of Intelligence Collection," speech, Washington, D.C., July 18, 2013, http://www.dni.gov/index.php/newsroom/speeches-and-interviews/174-speeches-interviews-2009.

[b] PPD-28 added two additional restrictions: a requirement that the FISC approve the RAS and a reduction in the number of hops that can be followed from three to two (The White House, Presidential Policy Directive/PPD-28, "Signals Intelligence Activities," Office of the Press Secretary, January 17, 2014, http://www.whitehouse.gov/the-press-office/2014/01/17/presidential-policy-directive-signals-intelligence-activities).

[c] Testimony of Chris Inglis, Statement, House Permanent Select Committee on Intelligence, Hearing on "How Disclosed NSA Programs Protect Americans, and Why Disclosure Aids Our Enemies," June 18, 2013, http://icontherecord.tumblr.com/post/57812486681/hearing-of-the-house-permanent-select-committee-on.

According to the document mentioned immediately above, some of the most important FISA authorities include the following:

- Section 215 of the USA Patriot Act (corresponding to Section 501 of the FISA Act as amended), under which NSA collects information (metadata) about telephone calls to, from, or within the United States.
- Section 702, under which NSA is authorized to target non-U.S. persons who are reasonably believed to be located outside the United States but who are using U.S. communications service providers. NSA believes that collection under this authority is "the most significant tool in the NSA collection arsenal for the detection, identification, and disruption of terrorist threats to the U.S. and around the world."[25]
- Section 704, under which NSA is authorized to target a U.S. person outside the United States for foreign intelligence purposes if there is probable cause to believe the U.S. person is a foreign power or is an officer, employee, or agent of a foreign power. Use of this authority requires a specific, individual court order.
- Section 705(b), under which the Attorney General may approve a collection similar to that allowed under Section 704 against a U.S. person who is already the subject of a FISC order obtained pursuant to Section 105 or 304 of FISA.

In addition, PPD-28 limited the purposes for which SIGINT collected in bulk can be used to six purposes, namely for detecting and countering the following:[26]

> (1) Espionage and other threats and activities directed by foreign powers or their intelligence services against the United States and its interests;
> (2) Threats to the United States and its interests from terrorism;
> (3) Threats to the United States and its interests from the development, possession, proliferation, or use of weapons of mass destruction;
> (4) Cybersecurity threats;
> (5) Threats to U.S. or allied Armed Forces or other U.S. or allied personnel; and
> (6) Transnational criminal threats, including illicit finance and sanctions evasion related to the other purposes named in this section.

There have been two important changes to the Section 215 program as a result of the current public debate. A January 2014 presidential state-

---

[25] Joint Statement: NSA and Office of the Director of National Intelligence.
[26] The White House, Presidential Policy Directive/PPD-28, "Signals Intelligence Activities," 2014.

ment announced that the number of "hops" would be reduced from three to two and that the FISC would be tasked with approving RAS selectors.[27]

---

[27] See The White House, "Remarks by the President on Review of Signals Intelligence," 2014, and U.S. Foreign Intelligence Surveillance Court, "In Re Application of the Federal Bureau of Investigation for an Order Requiring the Production of Tangible Things. Order Granting the Government's Motion to Amend the Court's Primary Order Dated January 3, 2014," Docket No. BR 14-01, Washington, D.C., http://www.uscourts.gov/uscourts/courts/fisc/br14-01-order.pdf.

2

Basic Concepts

Broadly speaking, the intelligence function involves the collection, analysis, and dissemination of information to decision makers. Intelligence analysts use all available sources of such information to understand problems of interest to decision makers. These sources include human intelligence, imagery, and a variety of other kinds of intelligence in addition to signals intelligence (SIGINT). This report focuses on signals intelligence.

In general, the intelligence process starts with decisions by policy makers on the areas of national security interest for which intelligence will be useful. Some of these areas cover imminent or anticipated threats, while others pertain to strategic intelligence to develop an understanding of regions or organizations that might become threatening. Based on the priorities stated by decision makers, intelligence officials in the community identify specific collection methods and opportunities that are expected to yield useful information. These methods and opportunities interact with and support each other (much as the various elements in an ecosystem interact with each other), so that, for example, a piece of information from one method may cue collection with another method or may corroborate or support information derived from another.

The intelligence process seeks information about both tactical matters (i.e., specific dangerous persons, groups, or plots, such as known terrorist organizations or plans to bomb subways or investigations of recent bombings) and strategic matters (i.e., a broad picture of a threat, such as a country's plans to build nuclear weapons). Increasingly, this is not a sharp distinction, because context is often important to understanding a tactical threat, and tactical information is required to respond to strategic threats.

A characteristic of tactical investigations is often (although not always) a highly compressed timeline. For example, in investigating a bombing, investigators must work quickly to determine whether the bomb that just exploded is the first in a series.

2.1 A CONCEPTUAL MODEL OF THE SIGNALS INTELLIGENCE PROCESS

Signals intelligence is defined by the National Security Agency (NSA) to be "intelligence derived from electronic signals and systems used by foreign targets, such as communications systems, radars, and weapons systems."[1] In the modern world, distinctions between paper records and electronic recordings that may once have been technically meaningful are increasingly obsolete as all forms of information storage become electronic.

In this section, the committee presents a simplified conceptual model of the parts and functions of the SIGINT process, which is used for further discussion. In Chapter 3, "use cases," examples of the use of SIGINT data in plausible scenarios, are shown. The description below is primarily technical in nature. Constraints on SIGINT imposed by law, regulation, and policy are discussed in Section 1.3.

As with other forms of intelligence gathering, SIGINT is conducted in response to requirements for intelligence from policy makers. Priorities are established by different agencies in the policy community and are reviewed at least annually. Based on these priorities, agencies in the Intelligence Community (IC), including NSA, design and develop mechanisms for collecting information in different locations, information that will meet the wide variety of policy maker requirements. To the extent possible, collection mechanisms are consolidated for greater efficiency, both between the various intelligence agencies and within NSA, as the entity charged with SIGINT collection. Thus, a given collection mechanism may provide information that is useful for a variety of different topics. This process seeks to avoid the development and deployment of collection mechanisms individually for each and every target, an approach that would be inefficient and expensive.

The committee's conceptual model of the SIGINT process is depicted in Figure 2.1. In this model, NSA extracts signals data from various sources, filters it for items of interest, stores the items, analyzes them, and disseminates selected information to policy makers and other units of the IC. (Not described here, and discussed later in the report, are the

---

[1] See NSA, "Signals Intelligence," http://www.nsa.gov/sigint/, accessed January 16, 2015.

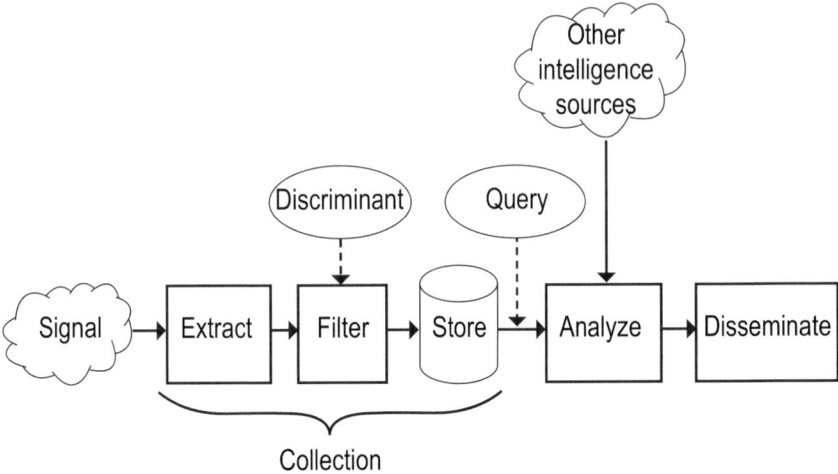

FIGURE 2.1 A conceptual model of signals intelligence.

audits and other measures to establish compliance with rules and regulations concerning personal privacy.) There are many signal types; among the most important are the digital signals that carry the voice content of telephone calls. Information pertaining to telephony is also collected as SIGINT; this is information about the calling and called telephone numbers and time and duration of call—so-called "telephony metadata." Internet communications, such as email or commands to search engines, may also be collected, and, once again, a distinction is drawn between content and metadata.

2.1.1 Collection

Signals are derived from many sources, but the specific steps taken to winnow large data streams to those that are manageable and potentially productive are the same regardless of the source. Figure 2.1 shows how one signal might be collected. The first three steps in the SIGINT model, taken together, are what the committee informally calls *collection*:[2]

---

[2] The committee's definition of *collection* differs from that used by NSA in certain ways. See, for example, NSA, *NSA's Civil Liberties and Privacy Protections for Targeted SIGINT Activities Under Executive Order 12333*, NSA Director of Civil Liberties and Privacy Office Report, October 7, 2014, https://www.nsa.gov/civil_liberties/_files/nsa_clpo_report_targeted_EO12333.pdf. See also footnote 3 in this chapter.

- *Extract.* The first step is to obtain the signal from a source, convert it into a digital stream, and parse the stream to extract the kind of information being sought, such as an email message or the digital audio of a telephone call. Extraction interprets layers of communications and Internet protocols, such as Optical Transport Network (OTN), Synchronous Digital Hierarchy (SDH), Ethernet, Internet Protocol (IP), Transmission Control Protocol (TCP), Simple Mail Transport Protocol (SMTP), or Hypertext Transport Protocol (HTTP). In cases where business records are sought, this step extracts and reformats relevant SIGINT data from a business record format used by the business.
- *Filter.* This step selects, from all the items extracted, items of interest that should be retained. It is sometimes controlled by a "discriminant," which the IC agency running the collection provides to describe in precise terms the properties of an item that should be retained. For example, a discriminant might specify "all telephone calls from 301-555-1212 to Somalia," "all telephone calls from France to Yemen," or "all search-engine queries containing the word 'sarin.'" If there is no discriminant, then all extracted items are retained.
- *Store.* Retained items are stored in a database operated by the U.S. government. This is the point at which collection is deemed in this model to occur for the retained data.[3] By contrast, the previous steps are fleeting, with data processed in near real-time (keeping data only for short periods of time—minutes to hours—for technical reasons) as fast as it is supplied, with all but the items to be retained discarded. Items collected from separate sources are usually combined into a modest number of large databases to facilitate searching and analysis.

In modern communication systems, traffic from many sources and destinations is aggregated into a single channel. For example, the radio signals to and from a base station serving all mobile phones in a cell are all on the same radio channels, and all of the IP packets between two routers may be carried on the same fiber. With rare exceptions, there is no single physical access point comparable to the central office connection of a landline telephone at which to observe only the items of interest and nothing more. Reflecting this reality, the committee's definition of "collection" says that SIGINT data is collected only when it is stored, *not* when it is extracted. Put another way, every piece of data that passes by a potential monitoring point must be machine-filtered as part of the

---

[3] Not everyone agrees on a definition of the word *collection*, which is widely used in policy, law, and regulation pertaining to SIGINT. This lack of collective agreement extends to entities within the IC itself. Moreover, subtle distinctions among the definitions lead to different views on certain SIGINT properties, especially its intrusion on privacy.

extraction process to determine whether it is potentially relevant or can be thrown away without further examination.

The committee notes that there are at least two differing conceptions of privacy with respect to when data are acquired. One view asserts that a violation of privacy occurs when the electronic signal is first captured, irrespective of what happens to the signal after that point. Another view asserts that processing the signal only to determine if it is irrelevant does not compromise privacy rights in any way, even if that signal is held for a non-zero period of time. In a technological environment in which different communications streams are mixed together on the same physical channel, picking out the sought-after communication stream *requires* the latter approach. Further, note that the committee has made a technical judgment about a useful definition of collection while remaining silent about what does or does not constitute an appropriate definition of privacy.

The committee also uses "collection" as a term to describe only *government* retention of data. If non-government actors acquire information from or about various parties in some legal manner but the government does not have access to that information, the government is not engaging in collection as a result of the actions of those parties. In contrast, if the government gains access to that information through technical or legal means and stores some or all of it for government use, it is reasonable to consider this collection.

Note that intelligence agencies narrow their focus throughout the various steps of collection as much as possible, both to comply with rules about what is allowed and to use their limited resources efficiently. Privacy protections of different sorts are applied at various points throughout the process. These include choices about where to extract signals and what discriminants to use, minimization procedures used to protect information about U.S. persons, and controls on how collected information can be used.

Notwithstanding the operation of the predecessor program to Foreign Intelligence Surveillance Act (FISA) Section 215, outside of the requirements of FISA, most agree now that the IC can target U.S. persons only when permitted explicitly with Foreign Intelligence Surveillance Court (FISC) involvement using procedures designed to ensure Fourth Amendment protections. The legal protections provided by the Fourth Amendment and various domestic legislation, such as FISA, distinguish between foreign and U.S. persons; in particular, the latter enjoy the protections of the Fourth Amendment. In cases where information about U.S. persons is collected as a part of authorized foreign intelligence collection activities, minimization rules approved by the U.S. Attorney General require special handling for privacy protection, consistent with foreign intelligence needs, which typically will require removing the names of U.S. persons or other

BASIC CONCEPTS   31

TABLE 2.1 Hypothetical Call Detail Records as They Might Appear in a Signals Intelligence Database

Caller	Called	Call Start Time	Call Duration
+1-617-555-0131	+1-703-555-0198	2014:10:3:15:45:10	3:41
+1-703-555-0198	+1-703-555-0013	2014:10:3:15:49:10	1:10
+1-415-555-0103	+963 99 2210403	2014:10:3:16:01:43	73:43
+1-603-555-0141	+1-603-555-0152	2014:10:3:22:10:03	3:01
+1-617-555-0183	+1-413-555-0137	2014:10:3:22:33:48	7:03
+1-802-555-0141	+1-802-555-0108	2014:10:3:22:41:17	3:02

NOTE: In this hypothetical example of call detail records as they might appear in a signals intelligence database, the call shown in the first line might be relaying a message through an intermediary at +1-703-555-0198. The call on the third line is to an international number, which might belong to a foreign national or a U.S. person. The call in the fourth line was probably ordering a pizza, since a directory of telephone numbers reveals that the called number is a pizza shop.

identifying information prior to dissemination. Of course, the names of U.S. persons can be included when necessary to understand the foreign intelligence information.[4]

Stated policy calls for strict rules for the dissemination of identities of U.S. persons in intelligence reports.

2.1.2 Analysis

Intelligence collection results in large databases holding records that are expected to have intelligence value. (Table 2.1 provides a hypothetical example of records in such a database.) In counterterrorism investigations, an analyst generally starts with a "seed," an identifier of a communications endpoint that has been obtained in the course of intelligence gathering and is deemed relevant to a possible threat. The analyst uses the seed identifier to formulate one or more queries of the databases to seek more information, for example, identifiers for other parties communicating with the seed. The analyst may also query for communications content, if it exists or can be obtained. Thus, analysts can build a pattern of a seed's connections to other parties and/or to other data that provide a richer and fuller picture of that party's role within a larger enterprise, such as a terrorist organization. Other databases may be consulted as

---

[4] The committee's understanding, based on the briefings it received, is that most data incidentally collected about U.S. persons are never examined, because U.S. person data is not returned in response to analyst queries for foreign intelligence information.

well. In this way, analysts can build a network that depicts how parties of interest relate to one another and characterize the activities of each of the parties in a network or more formally structured enterprise.

Analysts use a variety of software tools as they work with SIGINT data. They may use tools to formulate queries or display the results (e.g., see Figure 3.1). They may set up "standing queries" (which need special approval) that run each day to report new events associated with their active targets. Using results of queries of the data, they build a record of data and evidence for investigations in a "working store," a set of digital files separate from the SIGINT databases.

2.1.3 Dissemination

The last step in the SIGINT process is dissemination. SIGINT analysts will routinely disseminate the results of their work to others, both inside and outside the IC. For example, NSA analysts working on a specific terrorism investigation might disseminate their findings to other analysts and collectors who are working on related issues or directly to policy makers who may choose to take action based on the SIGINT.

Like the initial collection, SIGINT dissemination is governed by various laws and regulations designed to protect the sources and methods involved in the collection as well as the privacy and civil liberties of the subjects of the collection, especially if the intelligence involves U.S. persons.[5] Specifically to the latter, and pursuant to U.S. Signals Intelligence Direcive (USSID) 18,[6] such reports will normally cloak the identity of U.S. persons until a reader of the report specifically asks for the identity to be disclosed and provides a valid reason for the release, such as initiating a further investigation. This process is designed to ensure that both the requesting agency and NSA, as the disseminator of the information, can verify that disclosing this sensitive information is appropriate and necessary to understand the foreign intelligence value of the report.

2.2 BULK AND TARGETED COLLECTION

Presidential Policy Directive 28 (PPD-28) asks whether it is feasible to create software that could replace "bulk collection" with "targeted

---

[5] Section 4 of PPD-28 indicates that the IC should endeavor to give the same protections to foreign persons as to U.S. persons with regard to the retention and dissemination of identifying information.

[6] National Security Agency, "United States Signals Intelligence Directive USSID SP0018, (U) Legal Compliance and U.S. Persons Minimization Procedures," Issue Date January 25, 2011, approved for release on November 13, 2013, referred to as USSID 18, http://www.dni.gov/files/documents/1118/CLEANEDFinal USSID SP0018.pdf.

collection."[7] This section attempts to explain this distinction, which, unfortunately, is quite unclear. This question will be answered in Chapter 4.

Bulk collection results in a database in which a significant portion of the data pertains to identifiers not relevant to current targets. Such items usually refer to parties that have not been, are not now, and will not become subjects of interest. Moreover, they are not closely linked to anyone of that sort: knowing to whom these parties talk will not help locate threats or develop more information about threats. Bulk collection occurs because it is usually impossible to determine at the time of filtering and collection that a party will have no intelligence value. Although the amount of information retained from bulk collection is often large, and often larger than the amount of information retained from targeted collection, it is not their size that makes them "bulk." Rather, it is the (larger) proportion of extra data beyond currently known targets that defines them.

Targeted collection tries to reduce, insofar as possible, items about parties with no past, present, or future intelligence value. This is achieved by using discriminants that narrowly select relevant items to store. For example, if the email address hardcase45@example.com was obtained from a terrorist's smartphone when he was arrested, using a discriminant to instruct the filter to save only "email to or from hardcase45@example.com" would result in a targeted collection. Some or many of the people communicating with this person might turn out to have no intelligence value, but the collection is far more selective than, say, collecting all email to or from anyone with an email address served by aol.com. A discriminant could be a top-level Internet domain, a country code (e.g., .cn for China, .fr for France), a date on which communication occurred, a device type, and so on. A discriminant could even refer to the content in a communication, such as "all email with the word 'nuclear' in it." Note that if a discriminant is broadly crafted, the filter may retain such a large proportion of data on people of no intelligence value that the collection cannot be called "targeted."

PPD-28 seeks ways to reduce or avoid bulk collection in order to increase privacy and civil liberty protections for those not relevant to the intelligence collection purposes. Note that there is no precise threshold in collecting data on such "harmless" persons that will distinguish between bulk and targeted; it's a matter of degree. Also note that the bulk/targeted distinction applies broadly to different data types: telephony content, metadata, business records, Internet searches, and so on.

The fundamental trade-off, which can be seen in the Chapter 3 "use cases" and is explored further in Chapter 4, is between more intrusive

---

[7] The White House, Presidential Policy Directive/PPD-28, "Signals Intelligence Activities," Office of the Press Secretary, January 17, 2014, http://www.whitehouse.gov/sites/default/files/docs/2014sigint_mem_ppd_rel.pdf.

information collection that may yield extremely valuable information about threats unknown at the time of collection and less intrusive information collection that may miss information about dangerous threats.

Bulk and targeted collection can apply to many different kinds of communication modalities—telephone, email, instant message, and so on. Various web-based applications, such as electronic banking or online shopping that allow users to exchange information electronically, are among these modalities, even if they are not usually thought of as means for communication, per se. Obtaining phone metadata under Section 215 authority also counts as bulk collection.

2.3 DEFINITIONS OF CRITICAL TERMS

The laws, regulations, court rulings, and other writings about SIGINT use a number of terms to describe intelligence gathering and analysis. These terms are not always used precisely or consistently. Intelligence and law enforcement cultures use different words for the same concept or the same word for slightly different concepts. It is easy, when describing and debating intelligence processes, to stumble over problems of definition rather than of substance. Indeed, for several years in some instances, NSA analysts were accessing the database of domestic telephone metadata without proper reasonable and articulable suspicion (RAS) authority; this was due to differing NSA and FISC definitions of the word "archive."[8] Several presenters to the committee acknowledged these problems and indicated that the IC is continuing to work on them. The term *target* is also used loosely and in different forms throughout the community.

The preceding section addresses the definitions of *bulk* and *targeted* collection; this section provides working definitions adopted by the committee for several other key terms. For the purposes of this report, the committee has formulated the following lexicon (see Figure 2.2):

- *Identifier:* A text or bit string that denotes a communication end point.
- *Unknown:* An identifier that may or may not have intelligence value.
- *Ruled out:* An identifier that has been determined to have no intelligence value at the present time.

---

[8] John DeLong testimony to committee; see also Memorandum of the United States in Response to the Court's Order Dated January [sic] 28, 2009 at 11, In re Production of Tangible Things From [REDACTED], No. BR 08-13 (FISA Ct. February 17, 2009), http://www.dni.gov/files/documents/section/pub_Dec%2012%202008%20Supplemental%20Opinions%20from%20the%20FISC.pdf.

BASIC CONCEPTS 35

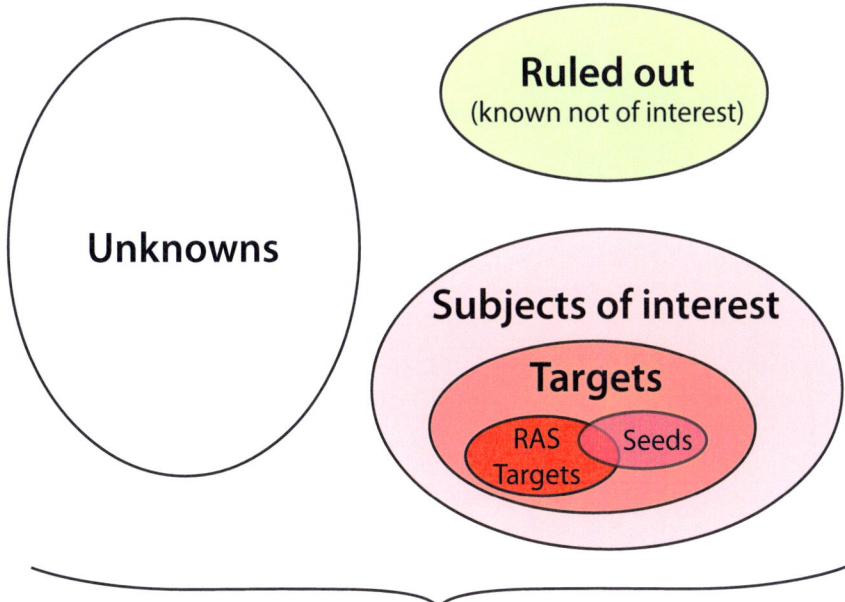

FIGURE 2.2 Classification of identifiers used in signals intelligence analysis.

- *Subject of interest:* An identifier that may have intelligence value and is likely to be part of an intelligence investigation.
- *Target:* A subject of interest that may be a security threat.
- *Seed:* A subject of interest that is used as the starting point for an intelligence investigation.
- *RAS target:* A target for which there is a reasonable and articulable suspicion that the person is associated with a foreign terrorist organization.[9]

For the purposes of this report only, and realizing that they may have different and possibly broader meanings in the IC, the committee uses the working definitions presented in Box 2.1, drawn from statutes and its understanding of IC practices in the context of SIGINT and technology. (For definitions used by the U.S. SIGINT System, see USSID SP0018, Section 9.[10])

---

[9] RAS is a term of art used in the context of Section 215 collection. See David Kris, On the bulk collection of tangible things, *Journal of National Security Law and Policy* 7:209, 2014.

[10] National Security Agency, USSID 18, approved for release in 2013.

BOX 2.1
Working Definitions in Signals Intelligence and Technology

identifier — A text or bit string that denotes a communication endpoint, such as a telephone number, mobile phone subscriber number, Internet Protocol (IP) address, or email address.

subject of interest — An identifier of a party (person, group) that may have intelligence value and is likely to be part of an intelligence investigation.

target (n, adj) — A subject of interest in an intelligence investigation.
This term is used liberally by the Intelligence Community (IC) to denote an identifier or person that is the subject of interest or surveillance.
A target need not be the principal subject of interest. For example, an associate of a known threat might be a target.
Note that a target can be a computer identified by its IP address.
Target identifiers may be used in selectors or discriminants to obtain, from a large collection of data, data pertaining only to the target.

seed (target) — An initial target used to start an intelligence investigation.

RAS target — A target for which there is a reasonable, articulable suspicion (RAS) that it is associated with a foreign terrorist organization. Foreign Intelligence Surveillance Act Section 215 requires a RAS target designation to permit certain queries.
In 2012, fewer than 300 identifiers met the RAS standards and were used as seeds in the Section 215 collection.[1]

query — Detailed instructions for searching a database of collected data. Note: this is consistent with computer technology usage, and is akin to an SQL query.
A query may have several "terms" or "selectors":
Example: "calls made from identifier x000325 after July 2, 2013"
Example: "Internet search requests using the term 'sarin' or emails containing 'poison gas'"

discriminant — Same meaning as query, but used in conjunction with filtering applied as part of collection. Discriminants must be simple enough to be applied in real time as signals intelligence (SIGINT) data is extracted and filtered.
This word appears explicitly in Presidential Policy Directive 28 (PPD-28) as part of the definition of "targeted collection."
Example: "all the email addresses used in communications to or from Yemen"

BASIC CONCEPTS 37

selector	(usually) A query term that cites a specific identifier. (sometimes) Any query term. Example: "calls made from identifier x000325" Example: "calls made from identifier x000235 or identifier y4576"
collection (of SIGINT data)	Storing SIGINT data on government-controlled information technology (IT) systems so as to enable authorized access by IC analysts and the software tools they use. Storing on a government-controlled IT system the results of a query of an external database constitutes collection in the sense of the committee's definition. Under the proposal to have communications carriers retain call detail records and allow authorized access to those records by IC analysts, the records at the carrier are not "collected" in the sense of the committee's definition, but if the records transmitted to the government in answer to a query are stored, they are considered collected.
bulk collection	Collection in which a significant portion of the retained data pertains to identifiers that are not targets at the time of collection. Note: Although the term "bulk" suggests that the set of collected data is large, and bulk data can indeed be large, size alone is not the controlling factor in defining bulk collection.
targeted collection	Collection that stores only the SIGINT data that remains after a filter discriminant removes most non-target data.
minimization	Procedures, approved by the Foreign Intelligence Surveillance Court (FISC), that must be "reasonably designed in light of the purpose and technique of the particular surveillance, to minimize the acquisition and retention, and prohibit the dissemination, of nonpublicly available information concerning unconsenting U.S. persons consistent with the need of the United States to obtain, produce, and disseminate foreign intelligence information." See, e.g., 50 U.S.C. §§ 1801(h)(1) and 1821(4)(A); USSID-18.[2]

continued

BOX 2.1 Continued

metadata	Loosely, "data about data," distinct from the data itself ("contents"). Sometimes called "non-content data."[3] There is no standard definition that enumerates metadata elements associated with a telephone call or an email transmission; instead, statutes and court orders that authorize collection of metadata list explicitly the elements that can be collected. However, metadata does not include "content of any telephone call, or the names, addresses, or financial information of any party to a call."[4] In telephony, generally includes calling and called numbers, duration of call, time of the call, and perhaps more information. In email, generally includes from and to email addresses, time of sending, IP addresses of email services, and the like. The "subject" and "re" fields of email headers are considered content, not metadata.[5]
call detail record (CDR)	Business records kept by telephone service providers (usually for billing purposes) detailing for each call information such as the calling and called numbers, the time and duration of the call, and possibly additional information. Sometimes called "telephony metadata."
business records	"Tangible things, including books, records, papers, documents and other items." Further, the FISC has ruled that the electronic form of these records counts as a tangible thing whose production can be compelled under Section 215. See USA Patriot Act,[6] also "Administration White Paper: Bulk Collection of Telephony Metadata under Section 215 of the USA Patriot Act."[7]

foreign intelligence information	"Foreign intelligence means information related to the capabilities, intentions, or activities of foreign governments or elements thereof, foreign organizations, foreign persons, or international terrorists." (PPD-28 and Executive Order 12333). For details see Foreign Intelligence Surveillance Act of 1978.[8] Foreign intelligence collection priorities are set annually at the policy level.
U.S. person	"A citizen of the United States, an alien lawfully admitted for permanent residence, an ... association [of citizens and permanents residents] or a corporation which is incorporated in the United States ..." For details see Foreign Intelligence Surveillance Act of 1978.[9]

---

[1] "Administration White Paper: Bulk Collection of Telephony Metadata Under Section 215 of the USA Patriot Act," August 9, 2013, p. 4. Found various places online, including http://big.assets.huffingtonpost.com/Section215.pdf.

[2] National Security Agency, "United States Signals Intelligence Directive USSID SP0018, (U) Legal Compliance and U.S. Persons Minimization Procedures," Issue Date January 25, 2011, approved for release on November 13, 2013, referred to as USSID 18, http://www.dni.gov/files/documents/1118/CLEANEDFinal USSID SP0018.pdf.

[3] See U.S. Department of Justice, Justice News, "Acting Assistant Attorney General Elana Tyrangiel Testifies Before the U.S. House Judiciary Subcommittee on Crime, Terrorism, Homeland Security, and Investigations," March 19, 2013, http://www.justice.gov/iso/opa/doj/speeches/2013/olp-speech-1303191.html.

[4] "Administration White Paper," 2014.

[5] FISC ruling, U.S. Foreign Intelligence Surveillance Court, "In Re Motion of Propublica, Inc. for the Release of Court Records, Docket No.: Misc. 13-09, The United States' Opposition to the Motion of Propublica, Inc. for the Release of Court Records," http://www.dni.gov/files/documents/1118/CLEANEDPRTT%201.pdf, p. 11.

[6] USA Patriot Act 2001, http://www.gpo.gov/fdsys/pkg/PLAW-107publ56/pdf/PLAW-107publ56.pdf, p. 17.

[7] "Administration White Paper," 2014.

[8] Ibid., p. 2, item (e).

[9] Ibid., p. 4, item (i).

3

Use Cases and Use Case Categories

To understand the uses of signals intelligence (SIGINT) data, several "use cases" are presented below. These are hypothetical scenarios that describe episodes in which analysts query SIGINT metadata as part of an investigation of a threat to national security, such as counterterrorism, or to stem the proliferation of weapons of mass destruction.[1] (Some public sources of information on actual cases are listed in Box 3.1.) The committee asked National Security Agency (NSA) briefers for unclassified use cases illustrating the use of metadata, under any authority, whether collected in bulk or targeted, whether foreign or domestic. The committee focused on how metadata are used, not on the authorities or restrictions under which it is collected. Three categories of use cases are presented below, which, the committee was told, account for the majority of metadata use: contact chaining, finding alternate identifiers, and triage.[2] This set is *not*, however, exhaustive.

The examples contain more detail than is strictly necessary to illustrate the use case categories. The detail is presented to show that an investigation may

---

[1] For scenarios of four counterterrorism investigations studied by the Privacy and Civil Liberties Oversight Board, see *Report on the Telephone Records Program Conducted under Section 215 of the USA PATRIOT Act and on the Operations of the Foreign Intelligence Surveillance Court*, January 23, 2014, http://www.pclob.gov/library/215-Report_on_the_Telephone_Records_Program.pdf, p. 144 ff.

[2] National Security Agency, presentation to the committee on August 28, 2014.

> **BOX 3.1**
> **Some Specific Cases of Signals Intelligence in Use**
>
> Very little has been made public about actual cases where U.S. signals intelligence has contributed to counterterrorism. A principal reason is that the Intelligence Community (IC) carefully protects information about sources and methods from adversaries. Nevertheless, information on some cases can be found in public speeches and testimony to Congress by IC leaders and in two reports prepared by the Privacy and Civil Liberties Oversight Board.
>
> The accounts of these cases are incomplete and possibly inconsistent. The selection of the cases that were made public, the details of the accounts, and their significance have all been controversial.
>
> Pointers to some of this public information are provided below, not because the committee endorses the views of its authors, but simply to supplement the abstract use case categories presented in this chapter with some concrete examples:
>
> - Testimony by Gen. Keith Alexander and others before the House Select Committee on Intelligence, June 18, 2013, http://icontherecord.tumblr.com/post/57812486681/hearing-of-the-house-permanent-select-committee-on.
> - Four cases using Foreign Intelligence Surveillance Act (FISA) Section 215 authority:
>   —Basaaly Moalin, financial support of Al Shabab.
>   —Najibullah Zazi, plotted to bomb the New York Subway system.
>   —David Coleman Headley, helped plan the 2008 Mumbai attack.
>   —Khalid Ouazzani, suspected of plotting to bomb the New York Stock Exchange.
> - Described in Privacy and Civil Liberties Oversight Board, *Report on the Telephone Records Program Conducted under Section 215 of the USA Patriot Act and on the Operations of the Foreign Intelligence Surveillance Court*, http://www.pclob.gov/library/215-Report_on_the_Telephone_Records_Program.pdf, p. 144 ff.
> - Some uses of FISA Section 702 authority are described in Privacy and Civil Liberties Oversight Board, *Report on the Surveillance Program Operated Pursuant to Section 702 of the Foreign Intelligence Surveillance Act*, http://www.pclob.gov/library/702-Report.pdf, p. 104 ff.

- Depend on different kinds of data,
- Use different analysis techniques or use common techniques in different ways,
- Use both bulk and targeted SIGINT collection, and
- Expect to reveal U.S. persons, whose constitutional rights must be protected.

Note that the SIGINT data used in these examples are metadata collected from telephone and email communications. The only metadata elements used are the "to" and "from" identifiers in the form of telephone numbers or email addresses, or the Internet Protocol (IP) address of a computer used for communication. Collection methods are not described, and it is assumed that the data are collected in such a way that they contain the entries that are required to satisfy the scenario. The Intelligence Community (IC) may collect additional kinds of SIGINT metadata.

3.1 CONTACT CHAINING

Communications metadata, domestic and foreign, are used to develop contact chains by starting with a target and using metadata records to indicate who has communicated directly with the target (one hop), who has in turn communicated with those people (two hops), and so on. Studying contact chains can help identify members of a network of people who may be working together; if one is known or suspected to be a terrorist, it becomes important to inspect others with whom that individual is in contact who may be members of a terrorist network. Similarly, studying contact chains can help analysts to understand the structure of an organization under investigation.

3.1.1 Use Case 1

In Use Case 1, the U.S. government has identified a Somali pirate network that includes target A. An analyst queries and displays all the call contacts to or from A's telephone number in the last 18 days. Some contacts are identified as already known targets; others are undetermined. The analyst invokes a similar query and display for target B, who has communicated frequently with A, and notes that there are three people, not yet determined to be targets, who have been in contact with both A and B. The analyst can see this relationship immediately, because the contact sets of A and B are displayed as a network, with contacts as nodes, linked by lines to indicate calls. The analyst invokes the query-and-display function again on one of these three, C, and discovers this person is in contact not only with targets A and B but also with other known pirates. Perhaps C is a "missing link" between the networks in which A and B are operating.[3]

Many contacts uncovered this way are ruled out as having no intelligence value. Calls to a car mechanic, an IT help desk, or an automated

---

[3] "Inside the NSA," *60 Minutes*, CBS News, video segment, December 15, 2013, 3:40-4:45, http://www.cbsnews.com/videos/inside-the-nsa/. The transcript for the *60 Minutes* segment is at CBS News, "NSA Speaks out on Snowden, Spying," December 15, 2013, http://www.cbsnews.com/news/nsa-speaks-out-on-snowden-spying/. Note that the video that plays on the page with the transcript is not guaranteed to be the correct segment of *60 Minutes*; the URL for the correct video segment is given above.

weather report are likely to be ruled out, although perhaps some may later be found to have intelligence value. Further, laws or regulations restrict what an analyst is allowed to do. For instance, there are special rules applied to subjects of interest who are or might be U.S. persons and various (and differing) sets of rules depending on which authority allowed the collection of the underlying information (see Section 1.4).

3.1.2 How Metadata Are Used in Contact Chaining

Either bulk or targeted collection can lead to the result in Figure 3.1. Since A and B are targets, targeted collection using a discriminant that specifies "collect all calls to or from A or B" would collect all the contacts and subjects shown in the figure. However, if all calls between A or B and C occurred before either A or B was identified as a target, later collection targeted on A or B will not find C by way of A or B, but might find C because of communication with some other target. Bulk collection provides useful "history," because it does not limit collection to only the targets known at the time of collection.

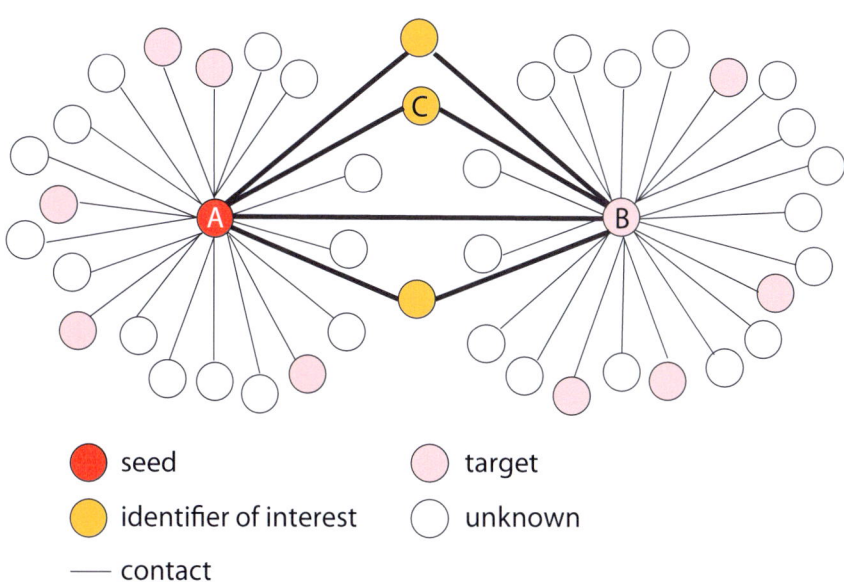

FIGURE 3.1 A network of contacts among identifiers.

3.2 FINDING ALTERNATE IDENTIFIERS

Targets may use several communication channels, each characterized by a specific identifier—in the example above, a telephone number, email address, or IP address. Targets may use different channels as a matter of convenience or as a form of operational security to try to evade detection by spreading their communications over several channels, by initiating new channels, or by stopping use of some channels. In some cases, identifiers may be assigned by the technology, such as an Internet service provider (ISP) that assigns a temporary IP address to a laptop.

An analyst can continue tracking a target by knowing the set of identifiers the target uses and tracking changes to the set over time, for example, when the target switches to a different email address. Activity detected using these identifiers is an important part of intelligence about the target. For example, a frequently used identifier that goes silent or that is found to have moved (e.g., by being detected at a different site) may indicate target activities of interest.

To succeed, alternate identifiers must be found quickly, with a speed and rate that meets or exceeds that with which the target acts. If targets are changing phone or email identifiers every day, the surveillance required to track the changes must be undertaken at a similar rate.

3.2.1 Use Case 2

In Use Case 2, an international cyber-criminal is thwarted when U.S. government access to his email communication allows anticipation and mitigation of a cyber attack. In response, the criminal transfers his communications to an alternate identifier—using a smaller ISP within the United States that is known for outspoken resistance to government surveillance. The U.S. government, via a Foreign Intelligence Surveillance Court (FISC) order, obtains bulk email metadata from the ISP, also imposing a gag order on the ISP and preventing deliberate or inadvertent disclosure of surveillance to the cyber-criminal. The intent of this action is to uncover the criminal's alternate identifier for the purpose of collecting additional intelligence.

The alternate identifier technique is applied to the email metadata obtained from the ISP, in order to find a new identifier that communicates with the same identifiers as the old email address, leading to the discovery of an alternate identifier used by the cyber-criminal.

In the following use case, alternate identifiers are used in a more complex scenario that combines communications surveillance with other intelligence methods.

3.2.2 Use Case 3

In Use Case 3, country X is a U.S. adversary that produces chemical and biological warfare weapons. The U.S. policy community wants continued monitoring of the program and to know if the country is supplying terrorists with weapons of mass destruction (WMD). The IC knows the following about the program:

- It is run under the cover of a medical research institute at the major university in the country. The institute also conducts legitimate medical and pharmacological research. There are also a variety of known and suspected laboratories associated with the program spread throughout the country.
- The institute's doctors, scientists, and researchers were trained in Europe, Russia, and the United States. The institute seeks medical research equipment from legitimate suppliers around the world.
- Plague, anthrax, and malaria are endemic to the country. The institute regularly works with the United Nations and international aid organizations to mitigate the threat posed by these and other diseases.
- Some working with the institute have provided "medical aid" to the Sons of the Western Sun, a U.S.-designated terrorist group attempting to overthrow the government in a neighboring country.

The IC goals are to

- Identify equipment and materials that the institute or its associated laboratories are attempting to purchase and who the suppliers are or could reasonably be.
- Locate and identify all the laboratories and facilities in the country associated with the institute.
- Determine research topics being pursued by members of the institute.
- Track communication between Sons of the Western Sun and members of the institute.
- Determine the view on WMD of the country's leadership and the directions provided to them by the institute.

To obtain information relevant to these goals, the IC may collect against the institute, the country, and the terrorist organization. The collection options are constrained by the following:

- Only persons considered loyal are allowed to travel overseas. They are also very wary of communications intelligence activities.

- Few foreigners travel to the country. The U.S. Embassy is heavily watched, and the staff is small.
- The institute regularly buys material and equipment online and often will contact suppliers with unsolicited emails asking for information on a wide range of products and services.
- Almost all telephonic communications is by cell phone. Twitter is a national pastime.
- The Sons of the Western Sun are believed to obtain substantial financial, logistical, and personnel support from elements in Europe and the United States, many of whom are unknown.

Use Case 3 illustrates a complex scenario in which several different ways of gathering intelligence may be involved. Most likely, the entire institute would be the focus of communications data collection. According to the committee's definition, this might be considered bulk collection, because it would collect a significant amount of data about communications of legitimate researchers who have no role in WMD efforts. However, focusing collection on the institute is less intrusive than collecting on the whole country. The alternate identifier method may be applied to the identifiers of everyone in the institute in order to track all ongoing communications. Correlations with known members of Sons of the Western Sun may help distinguish targets from innocents.

3.2.3 Use Case 4

In Use Case 4, following the events of Use Case 3, country X eliminated its WMD programs. The United States aided with the destruction of the weapons, but despite public declarations, the IC remains convinced that a number of facilities were never identified by the country. The new government has been rumored in press articles to want to re-establish the WMD program. U.S. policy makers are concerned about a new arms race in the region and want to know the status and intention of the country toward its WMD program. The IC knows the following:

- Because of the thaw in U.S.-X relations, the scientists who worked on the old WMD program have been traveling widely in the United States and Asia.
- Large numbers of citizens of X currently travel freely between X and the United States, and a number of U.S. tourists travel yearly to X to bathe in the renowned hot springs.
- Several key proliferators associated with the old WMD program have winter villas in X. They were known to buy goods from both U.S. and Asian suppliers for the program.

• The government of X and most of its leading citizens have bank accounts in the United States, among other places.

Analysts have to determine the current status of the WMD program and leadership intentions toward the program. Unfortunately, after X agreed to dismantle its WMD program, most collection efforts on the program were ended or drastically reduced. Thus, the IC goals are to

• Determine if old proliferators started shopping for WMD-related materials and equipment.
• Identify the actual intention of the government of X or other senior policy makers toward the WMD program, despite public pronouncements.
• Identify current activities at all previously known WMD sites and possible new facilities, and identify the use of new agents to purchase WMD-related materials and equipment.
• If a WMD program is identified, determine the what, where, who, and how.

This example draws on many intelligence sources of which SIGINT is only one. Part of the approach will be to collect bulk communications data focused on areas where IC analysts expect the scientists to be communicating. They may use some of the same identifiers they used before the WMD program shutdown. Doubtless, new identifiers will come to light, some identifying U.S. persons and, therefore, requiring minimization procedures.

Intelligence reports from the period *before* the shutdown may contain references to identifiers or other evidence that will help target or focus collection or seek contemporaneous alternate identifiers. It is possible that during the shutdown a reduced collection effort was sustained in order to monitor the termination of WMD activity; parts of that data that have not exceeded the IC's retention limits may be used in the alternate identifier search.

3.2.4 How Metadata Are Used in Finding Alternate Identifiers

Finding alternate identifiers depends on collecting timely communications metadata in bulk. The collection must be in bulk in order for the metadata to include the new identifiers, which are not known at the time of collection to be associated with a target because they are *two hops* from the old identifier. Collection that is focused around the target (i.e., communications channels and modes that the target is known to use) is used, in part, to limit intrusion on innocents. Focus is also driven by concerns of cost and computer processing time to run the correlation algorithms.[4]

Note that the two-hop restriction announced by the President in January 2014 on queries of domestic telephone metadata still allows alternative identifiers to be found for known domestic reasonable and articulable suspicion (RAS) targets.[5] Starting from a target, a query will find all the target's one-hop contacts, then find all two-hop contacts; among these, there may be an identifier that communicates with many of the target's one-hop contacts. This identifier may be an alternate identifier for the target.

Reverse targeting is another approach to find a target's alternate identifiers by working backwards from persons known to be in contact with the target. In this approach, each identifier that communicates with the target is used as a query against bulk metadata or as a selector in future targeted collection, which will reveal any new identifier that communicates with the target's previous communication partners. Because this method explicitly collects against persons known *not* to be persons of interest, apart from the fact that they communicate with the target, the use of this method raises extra privacy concerns. Current policy and statutes forbid the use of this method under certain authorities.[6]

3.3 TRIAGE

Investigations or events may uncover lists of identifiers that need triage—that is, categorizing identifiers according to the danger that their owners might pose to national security. Queries about each identifier are made to the IC's databases to determine whether the identifier can be matched against a currently known target, is related to a target, or exhibits other properties of a dangerous person. Most often, the list presented for triage cites only identifiers and not names of persons.

Queries for triage are matched against historical metadata (both bulk and targeted) to find evidence of connections between the identifier and events or people that were or are of interest. The identifier may have once been a target, but the information about the target has since been discarded. Or the identifier may have been retained because it arose in the

---

[4] Note that these algorithms will examine metadata associated with many innocents. The set of identifiers examined will certainly include all those associated with reverse targeting, explained below.

[5] The White House, "Remarks by the President on Review of Signals Intelligence," Office of the Press Secretary, January 17, 2014, http://www.whitehouse.gov/the-press-office/2014/01/17/remarks-president-review-signals-intelligence.

[6] See U.S. Code Title 50, Chapter 36, Subchapter VI, § 1881a, Procedures for targeting certain persons outside the United States other than United States persons.

course of an investigation. In any case, the alternate identifier technique is used to find new identifiers related to the one offered for triage that might be used by the same person.

3.3.1 Use Case 4—Extension of the Scenario

In Use Case 4, when the IC resumes investigation of country X, the many identifiers obtained during the original investigation must be triaged for use in the new context.

3.3.2 Use Case 5—The Immediate Response After a Terrorist Incident[7]

In Use Case 5, a bombing suspect is identified, along with an associated email address. The triage process, which includes finding alternate identifiers used by the suspect, is used to quickly find possible associates of the suspect and other information about these identifiers held in IC databases.

3.3.3 How Metadata Are Used in Triage

Triage benefits from all retained metadata, bulk or targeted, because it seeks information about an identifier that may not have previously been the subject of explicit IC attention. A timely response in Use Case 5 requires historical information; initiating targeted collection using the relevant email addresses obtained only after a suspect was identified would mean that only future information would be available, probably trickling in too slowly to be useful. Other communications channels used by the suspect in the past, such as a cell phone, might never be found using present and future SIGINT data alone.

If an identifier presented for triage has never been associated with a target, then past bulk data is probably more likely to find direct associates, and contact chaining or alternate identifier techniques may lead to additional associates. But the identifier may also be found in metadata targeted to a particular terrorist organization, even though the identifier was not previously known to be associated with the organization.

3.4 CONCLUSION

The use case categories described above—contact chaining, alternate identifiers, and triage—illustrate the growing importance in intelligence work of discovering and defining networks. Indeed, doing so may be as

---

[7] National Security Agency, presentation to the committee on August 28, 2014.

important as examining content to know what members of the network are saying. It is a task where bulk collection is especially important.

The use case categories above do not exhaust the use cases for SIGINT data, or even for telephony metadata. Many uses are ad hoc and do not fit into neat categories. For example, to find whether any of a number of targets associated with a terrorist group have communicated with any of a number of explosives suppliers, a query listing the targets and the suppliers can be constructed and applied against stored metadata. This is an iterative use of a simple query to determine whether one group contacted another.

The committee had hoped that analyzing use cases might suggest alternatives to bulk collection. But this path is limited, for two reasons: (1) the three categories do not cover all uses of SIGINT data and (2) the use cases show that both bulk and targeted collection are used. If the use case is focused in some way, targeted collection may provide enough data, especially if the focus has been under targeted surveillance for some time. For investigations that have little or no prior targeting history, bulk collection may be the only source of useful information. Thus it appears that it is the *context* of the investigation, rather than the *technique* for using collected metadata, that most influences the value of bulk collection.

4

Bulk Collection

This chapter builds on the use cases presented in Chapter 3 to describe in more general terms some ways in which bulk collection is used by the Intelligence Community (IC) and some of the challenges associated with alternatives that use targeted collection.

4.1 USES OF BULK COLLECTION

4.1.1 Information about the Past

If past events become interesting in the present for understanding new events, such as the discovery of a nuclear weapons test by a previously non-nuclear nation, historical facts and the context they provide will be available for analysis only if they were previously collected. Sometimes review of a targeted collection (e.g., against leaders of the non-nuclear nation) may reveal information for a new purpose that was not in mind when the information had been collected (e.g., the intent of the nation's leaders regarding nuclear weaponry). But sometimes useful information, such as the nexus of suppliers for the weapons technology, will be present only if previously there had been bulk collection. If it is possible to do targeted collection of similar events in the future, and they happen soon enough, then the past events might not be needed. If the past events are unique or if delay in obtaining results is unacceptable (perhaps because of press coverage or public demand), then the intelligence will not be as complete.

4.1.2 Tactical Intelligence

Chapter 3 presented several use cases illustrating the use of bulk collection for tactical intelligence. Tactical intelligence requires prompt attention to newly discovered targets and imminent threats. Collecting and saving information in bulk, without a specific set of targets, is the only way to have past information about a party on hand when that party becomes one of interest. Sometimes that information will be available because of a targeted collection in which certain uses were not yet realized. But sometimes information becomes interesting only because of new events or information, in which case previous bulk collection may be the only possible source. Targeted collection provides data only on present and future actions of parties of interest at the time of collection, but not on their past activities. For example, bulk collection may allow the identification of hostile actors and their associates because they made mistakes as their activities began, perhaps because of ineffective tradecraft or other casual interactions.

Understanding the significance of past activities and their actors is a feature of all investigations, foreign and domestic. In contrast to domestic law enforcement, however, the world of intelligence analysis has many fewer tools available for investigation. In hostile foreign environments, personal interviews and observations and records review are much more limited. Accordingly, the role of bulk data as a way to understand the significance of past events is important, and the loss of this tool becomes more serious. Of course, bulk collection can also be useful in a domestic context.

Some kinds of targeted collection are focused on topics rather than people, and some targeting based on topics will be more specific than others. For example, a discriminant that collects all queries to Internet search engines that ask about "sarin" or "poison gas" will collect information about many people of no intelligence interest because only a handful of those making such searches will be of actual interest. However, other discriminants citing specific military code names might yield information about fewer people who are of no intelligence interest.

4.1.3 Strategic Intelligence

In strategic intelligence, information is gathered to build understanding about a topic (e.g., climate change, migration patterns), an entity or area (e.g., region, nation, subnational group), or set of activities and sometimes takes the form of statistics or trends. Some examples include the following:

- Collecting against national, military, or organizational decision makers.
- Monitoring many types of communication among the officers in an army to help understand its morale, quality of training, or location. If the collection is only against the communications of army personnel, this might be considered a targeted collection.
- Bulk collection of communications can reveal health care, electric power, or agricultural data that is not reported accurately, or at all, by a government.
- Sampling everyday communications in a region can provide insight into local sentiment about political trends that might lead to, for example, a government overthrow. For example, social networking communications during the Arab Spring reported unfolding events in real time.

Some of the data collected for strategic intelligence is analyzed using statistical techniques: rather than looking for specific persons or groups, the goal is to monitor trends or patterns in communications that might lead to intelligence insights. This is one application of analytical techniques that are known today as "big data analytics."

4.1.4 Reference Data

Bulk collection is used to acquire reference data that supports other signals intelligence (SIGINT) collection or analysis. For example, analyzing communications data is greatly enhanced if analysts have "telephone directories" for organizations of intelligence interest—that is, a list of who's who in the organization and their communications identifiers.

Another role for bulk collection is to guide targeted collection; the IC refers to this role as "SIGINT development." For example, the decision about where to gather information can depend on knowing the target's likely modes of communication. Because the target will not assist the collector in this decision, the collector will have to discover the likely modes of communication—perhaps by collecting information from all the modes of communication that the target might use—to understand their significance for national security priorities. Similarly, the National Security Agency (NSA) may have the resources to thoroughly monitor only one of several communication channels, and learning that some of them carry mostly communications of U.S. persons would make those channels less likely to be selected because they are not apt to be good sources of foreign intelligence. In addition to making NSA's work more efficient, such decisions may reduce collection of information about people who are not of interest.

4.1.5 Increasing the Likelihood That Needed Information Is Available

Does bulk collection overwhelm analysts with too much data, as is sometimes argued? The "needle in the haystack" metaphor is relevant here. If the needle is not found in the smaller haystack, there are two approaches—not mutually exclusive—that may result in success. One approach is to add more hay (because that additional material may contain the needle of interest). A second approach is to do a smarter search (because a smarter search may turn up a needle that was in the haystack all along), such as using techniques described by Cortes et al.[1]

Of course, if the needle is not in the smaller haystack, no amount of smarter searching will help. The use case category of alternate identifiers illuminates this problem. An analyst has determined that a new target is of interest, where "new" means that this target has not previously been explicitly targeted for collection. With luck, previously targeted collection may provide information on alternate identifiers that the new target has used.

Adding bulk data may help, because, by definition, bulk collection may contain alternate identifiers. But there is still no guarantee, because the bulk data might have been collected in the wrong location or through the wrong communications channel, etc. The alternate identifiers might still be missed, even though they exist.

Is a smarter search more or less likely than the use of bulk data to result in identification of the needle? Without details of the specific use case in question, this question cannot be answered in the abstract. In practice, analysts do not know if the haystack contains the needle without analyzing all the data—so they cannot know when to stop adding more hay.

Thus, collecting more data is necessary but it is not necessarily sufficient. It is true that more data may burden the analyst, while increasing the risk of intruding on parties that are not of interest, and may still fail to provide the data of interest, even when such data exists. Still, if the necessary data is not already available, collecting more is the only possible way to find the needle. This trade-off between too much data and finding the necessary information is inevitable. Although it can sometimes be reduced, it cannot be eliminated.

4.2 ALTERNATIVES TO BULK COLLECTION

Below are some alternatives to present-day bulk collection practices that might mitigate some of the privacy and civil liberties concerns that

---

[1] C. Cortes, D. Pregibon, and C. Volinsky, Computational methods for dynamic graphs, *Journal of Computational and Graphical Statistics* 12(3):950-970, 2003.

such practices raise. Each also involves a variety of performance trade-offs when compared to bulk collection as currently handled.

• *Federating business record databases by allowing them to be held by telecommunications carriers and allowing authorized queries by the U.S. government.* This "federated storage" approach, which primarily applies to domestic collection, is discussed more fully in Chapter 5. By providing the U.S. government with certain access to business records stored by the telephone companies, this alternative retains the principal benefit of bulk collection by the U.S. government—access to telephone call history—but it is not as operationally effective as bulk collection. As detailed in Chapter 5, federation offers advantages for safeguarding privacy and enforcing policies. It also has disadvantages that include divergent incentives between the government and third parties, greater technical and organizational complexity, and potentially poorer performance.

• *Bulk analysis.* A class of alternatives extracts bulk SIGINT data from a source, applies "analysis algorithms" to all of it, saves the results of the algorithm, and then discards the SIGINT data. For example, one scheme might construct the contact network from call detail records (CDRs), store the entire network, and discard the CDRs. If a significant portion of the stored network pertains to nontargets, this technique should be viewed as a variant of bulk collection. Some proposals go even farther and use algorithms to fuse data from several different intelligence sources into an annotated "hypergraph," where the annotations retain information gleaned from intelligence data.[2] These schemes are arguably more intrusive on privacy and civil liberties than bulk collection of raw SIGINT, because they analyze and store a multi-source picture of many people who are of no intelligence value. Moreover, automatic analysis seems unlikely to replace human analysis, although it may be useful as an augmentation to what humans do.

• *Fast near-real-time targeting.* Targeted collection is most effective when targets can be added to the discriminant quickly as they are identified in previous communications. If a call from a target X to an unknown Y is rapidly followed by a call from Y, the second call may be significant—possibly a message being passed on. If the first call quickly adds Y as a new target in the collection discriminant, the second call will be collected; otherwise, it will not, because both ends of the call are unknown identi-

---

[2] J.C. Smart, Georgetown University, briefing to the committee on September 9, 2014; see also AvesTerra Program, "The FOUR-Color Framework: A Reference Architecture for Extreme-Scale Information Sharing and Analysis: Overview," V1.6, Georgetown University, October 2014, http://avesterra.georgetown.edu/sites/avesterra/files/4CF%20Overview%20%28V1.6%29.pdf.

fiers. Collection software could be designed to chain targets this way only if such chaining is pre-approved. While this approach may collect a few more rapidly unfolding scenarios, it does not provide the complete view of past events afforded by bulk collection.

- *Big data analytics.* It may be possible to use big data analytics to help narrow collection, even if the results from such analytical tools are not sufficiently precise to identify individual targets. That is, the government may be able to rely on the power of large private-sector databases, analytics, and machine learning to shape data collection constraints to data predicted to have high value. But even if the government collection becomes more narrowly targeted through the use of such analytic tools to develop the targeting, this is not necessarily a win for privacy. Depending on what aggregate data is used to determine the targeted government collection, use of such techniques may well raise privacy concerns. There will also be concerns that the methods used for targeting are akin to socially unacceptable profiling (e.g., targeting purchases of camping goods, males, ages 15 to 30). Thus, the use of big data analytics to provide better targeting may not be acceptable from a policy point of view, even if such techniques were to ultimately result in a more narrow government collection.

- *Cascaded filtering.* Some of these methods may benefit from the use of cascaded filtering. One benefit of this approach is that it allows one to reduce the computing burden by first applying cheap tests, followed by more expensive filters only if earlier filters warrant. For example, if metadata indicates a civilian telephone call to a military unit under surveillance, speech recognition and subsequent semantic analysis might be applied to the voice signal, resulting in an ultimate collection decision. Richer targeting may require enhancing the ability of collection hardware and software to apply complex discriminants to real-time signals feeds. Another benefit is that it will tend to reduce the amount of data that ends up being collected through fast and early filtering.

4.3 CONCLUSION

There is no doubt that bulk collection of SIGINT leaves many uncomfortable. Various courts have indeed questioned whether such collection is constitutional. This discomfort arises for many reasons. Some find the idea that the U.S. government collects vast amounts of communications signals information about unsuspected U.S. persons abhorrent to the very notion of democracy, while others object to this decision being made under the cover of secrecy.

This chapter has explored uses of bulk collection and technical alternatives the committee uncovered during its work that might mitigate some of the privacy and civil liberties concerns of that collection. None of these alternatives changes a fundamental point: A key value of bulk collection is its record of past SIGINT that may be relevant to subsequent investigations. If past events become interesting in the present because of new circumstances—such as the identification of a new target, indications that a nonnuclear nation is now pursuing the development of nuclear weapons, discovery that an individual is a terrorist, or emergence of new intelligence-gathering priorities—historical events and the data they provide will be available for analysis only if they were previously collected.

Conclusion 1. There is no software technique that will fully substitute for bulk collection where it is relied on to answer queries about the past after new targets become known.

This conclusion does not mean that all current bulk collection must continue. What it does mean is that a choice to eliminate all forms of bulk collection would have costs in intelligence capabilities. The analysis in this report provides a partial basis from which to make such policy choices.

Other groups, such as the President's Review Group on Intelligence and Communications Technologies and the Privacy and Civil Liberties Oversight Board have said that bulk collection of telephone metadata is not valuable enough to justify the loss in privacy.[3] This is a policy judgment, which is not in conflict with the committee's conclusion that there are no technical alternatives that can accomplish the same functions as bulk collection and serve as a complete substitute for it; there is no technological magic.

The committee was not asked to and did not consider whether the loss of effectiveness from reducing bulk collection would be too great, or whether the potential gain in privacy from adopting an alternative is worth the potential loss of intelligence information. Nor was it able to identify broad categories of use where substitution of alternatives might be possible or detect metrics that would inform such decisions. The Office of the Director of National Intelligence may wish to study these questions further.

Data retained from targeted SIGINT collection might be a partial substitute if the needed information was in fact collected. Bulk data held by other parties might substitute to some extent, but this relies on those

---

[3] President's Review Group on Intelligence and Communications Technologies, "Liberty and Security in a Changing World," http://www.whitehouse.gov/sites/default/files/docs/2013-12-12_rg_final_report.pdf, and Privacy and Civil Liberties Oversight Board, *Report on the Telephone Records Program Conducted under Section 215 of the USA PATRIOT Act and on the Operations of the Foreign Intelligence Surveillance Court*, January 23, 2014, http://www.pclob.gov/SiteAssets/Pages/default/PCLOB-Report-on-the-Telephone-Records-Program.pdf.

parties retaining the information until it is needed, as well as the ability of intelligence agencies to collect or access it in an efficient and timely fashion. Other intelligence sources and methods might also be able to supply some of the lost information, but the committee was not charged to and did not investigate the full range of alternatives that intelligence agencies could bring to bear. Note that all of these alternatives may introduce distinct privacy and civil liberties concerns.

Conclusion 1.1. Other sources of information might provide a partial substitute for bulk collection in some circumstances.

Because bulk collection cannot for practical reasons be truly comprehensive, it is itself inherently selective and unable to capture *all* relevant history. As a result, at least in some cases, it may be possible to develop techniques that would improve targeted collection to the point where it provides a viable substitute for bulk collection. Although such approaches might reduce the extent of collection against persons other than targets of interest, they might also introduce new privacy and civil liberties concerns about how such profiles are developed and used.

Rapidly updating discriminants of ongoing collections to include new targets as they are discovered will enable the collection of data that would otherwise be lost. If targeted collection can be done quickly and well enough, then there may be cases where information about past events becomes less important. But such an approach is not a substitute if the past events were unique or if the delay incurred in collecting the new information is unacceptable (because the threat is imminent or perhaps because of press or public demand for instant results).

Conclusion 1.2. New approaches to targeting might improve the relevance of the collected information to future use and would rely on capabilities such as creating and using profiles of potentially relevant targets, possibly by using other sources of information.

Chapter 6 describes some possibilities. Chapter 5 discusses technologies that can reduce risk and improve oversight and transparency.

5

Controlling Usage of Collected Data

5.1 WHY IT IS IMPORTANT TO CONTROL USAGE

Many people are concerned about how the large and rapidly growing amount of private data that exists online is handled, and whether privacy and civil liberties are properly protected. For signals intelligence (SIGINT), these concerns increase because the data is collected by the government. The disclosures by Edward Snowden have further increased concerns about the privacy of information that the National Security Agency (NSA) collects.

This chapter describes a number of ways to implement controls on the use of collected information. NSA is already using some of them, as the committee learned when the agency described in briefings how it complies with the legal authorities that govern its activities. NSA may be using other controls that the committee did not hear about, but there may also be opportunities to make compliance with the rules both more efficient and more transparent without increasing the compliance burden on analysts.

In understanding the security of any computer system, it is important to be clear about the "threat model," that is, the set of threats that the system must be defended against. In the context of bulk collection, there are three broad classes of threats:

- Entities outside the Intelligence Community (IC): hackers, cyber-criminals, foreign intelligence agencies;
- Lone insiders within the IC; and
- Misuse of the IC's capabilities, contrary to law or stated policies.

The last two are the main threats for most people concerned about privacy and civil liberties. Hence, the emphasis of this report is on controls, oversight, and transparency, which are the principal ways to address these threats.

5.2 CONTROLLING USAGE

Chapter 4 states the committee's conclusion that refraining entirely from bulk collection will reduce the nation's intelligence capability and that there is no kind of targeted collection that can fully substitute for all of today's bulk collection. However, the committee believes that controlling the *usage* of data collected in bulk (and indeed all data) is another way to protect the privacy of people who are not targets (see Figure 5.1).

Controls on usage can help reduce the conflicts between collection and privacy. There are two ways to control usage: manually and automatically. NSA automates some of its controls and plans additional automation. Despite rigorous auditing and oversight processes, however, it is hard to convince outside parties of their strength because necessary secrecy prevents them from observing the controls in action, and because popular descriptions of the controls are imprecise and sometimes wrong.[1] Examples of usage controls in place today are minimization (Section 1.4.1) and restricting queries to targets with reasonable and articulable suspicion (Section 1.4.3).

Technical means can *isolate* collected data and *restrict queries* that analysts can make, and the way these means work can be made public without revealing sources and methods.

This is similar to the well-established doctrine in cryptography[2] that the security of the system should depend only on keeping the cryptographic *key* secret, not on keeping the cryptographic *algorithm* secret. The main reason for this is that the algorithm exists in many more places than the key—with every sender or receiver of messages that uses the cryptosystem—so it is much harder to keep the algorithm secret and to change it if it is compromised. In contrast, a key is usually used only between a single sender and receiver, or at most a few of them, and only for a limited time, so it is much easier to keep it secret and to change it if it is compromised. In addition, a public algorithm may be more secure because many people can scrutinize it for weaknesses.

---

[1] See, for example, this newspaper account of President Obama's description of NSA practices: http://www.washingtonpost.com/world/national-security/obamas-restrictions-on-nsa-surveillance-rely-on-narrow-definition-of-spying/2014/01/17/2478cc02-7fcb-11e3-93c1-0e888170b723_story.html, *Washington Post*, January 17, 2014.

[2] First formulated by Kerckhoff in 1883. See Fabien Peticolas, *electronic version and English translation of "La cryptographie militaire."*

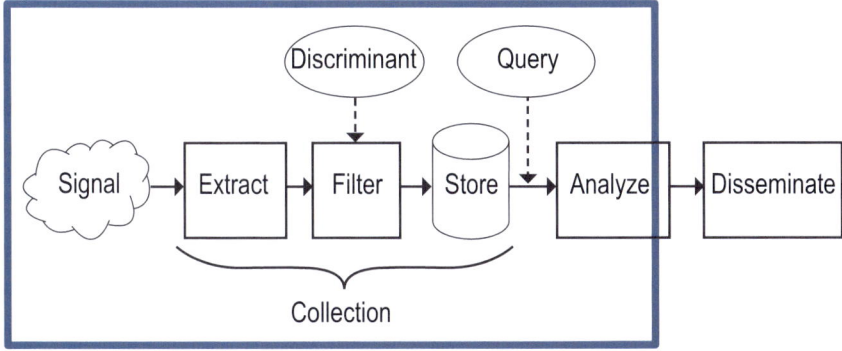

FIGURE 5.1 Focus of controls on usage.

In the same way, the specifics of actual use cases would be kept secret while the rules and the usage controls that enforce them are made public. This transparency makes the control of usage more credible.

Implementing usage controls in technology also forces those specifying the rules to be much more explicit than if they are providing instructions for human analysts to follow. Today, many of the descriptions for what is and is not allowed are in certain ways imprecise and ambiguous. Such ambiguities can lead to confusion and differing interpretations of the same rule. Furthermore, automatic controls may reduce the need for human labor implementing manual controls. Thus, technology may make the control more reliable and economical as well as more transparent.

It is impossible, however, for technical means to *guarantee* that information is not misused, because someone with properly authorized access can always misuse the information they obtain. This is like the "analog hole" in digital media; there are many ways to prevent digital copying, but when a human views or hears information, that information can be copied with a camera or sound recorder. Similarly, when an analyst sees information, he or she can misuse it. Thus misuse can only be *deterred* by the threat of punishment. Deterrence requires technical capabilities to detect access, to identify the (authorized) accessing party, and to audit records of access to spot suspicious patterns of access.[3] In addition, both

---

[3] Note that, to date, the only allegations that information collected in bulk has been used for an unauthorized purpose was the so-called "LOVINT" set of incidents in which some NSA analysts inappropriately used this data to track the activities of significant others.

manual and automatic controls are primarily aimed at analysts and others not in positions of authority. Detecting bad behavior by people in positions of authority needs multiple independent audit paths and oversight.

Lastly, it may be true that manual controls can be overridden more easily than automatic controls, because a technical change is usually more difficult than a procedural change. Changes are sometimes necessary to fix problems that arise, but whether it is good or bad for changes to be easily made is a policy judgment.

Manual and automatic methods can control usage in many ways, including the following:

- Constraining the selectors associated with targets to those that are approved in some way (e.g., analysts may target only those parties for which they have reasonable and articulable suspicion of involvement with terrorism);
- Limiting the time period for which data are accessible;
- Limiting the kinds of algorithms that are applied to data (e.g., algorithms that look for patterns, or various statistical techniques); and
- Using advanced information technology techniques to limit risk of disclosure, as described below.

The bulk of this chapter discusses how to control queries that analysts make against collected data. Controlling the use of such a large amount of data is critical, which is why the committee has emphasized it. When the data are queried, rules are applied about what uses of collected data are allowed. If a policy decision is made to continue bulk collection, protection of privacy and civil liberties will necessarily rely on these rules.

Once the results of a query are delivered to an analyst, other means must be used to control proper use of the data between queries and disseminated intelligence reports. These other means must be matched to what analysts actually do and to the tools they use. This cannot be done in the same way that queries on the collection database are controlled, for several reasons:

1. To do their jobs, analysts need flexibility to use the query results in many ways, such as combining them with other data or processing them with other programs, some perhaps written specifically for the current

---

These involved very few incidents (around a dozen). A letter from NSA to Senator Charles Grassley on September 11, 2013, details these incidents (see https://www.nsa.gov/public_info/press_room/2013/grassley_letter.pdf). According to testimony to the committee on August 23, 2014, by NSA Director of Compliance, the activities were uncovered through internal investigations.

purpose. These uses are much less standardized than the collection database and the ways of querying it, and it is not practical to control them in detail. The reason is that in order to construct software that tracks in detail the way that the inputs of a program affect its outputs, it is first necessary to formalize how the program works. This is usually much more difficult than writing the program in the first place.

2. Analysts share their work in progress with other analysts, so that even if the queries made by a single analyst return only 50 items, the queries made by 200 analysts may return 10,000 items altogether, and a single analyst or systems administrator may end up with all of these items.

3. In some cases, analysts import query results into commercial applications such as a spreadsheet like Excel or a statistical analysis system like SAS/STAT. It is not practical to modify these applications to track the way that their inputs affect their outputs, and it is impractical for the IC to develop its own substitutes.

4. Analysts do their work and store their data on workstations and servers that run commercial off-the-shelf operating systems because it is neither economical nor efficient for the IC to build its own operating systems and the applications. Furthermore, there are many versions of these systems in use at any given time, as is normal for any large organization. It is not practical to use these systems for fine-grained control of data.

It would be naive for the committee to claim that it understands what happens today, and presumptuous to pretend to design an ideal system for NSA's use. Furthermore, it is not enough to understand the normal information flow; possible changes, mistakes, and errors also need to be dealt with. For instance, something might change that would make yesterday's legitimate query unacceptable today. A target might have become a non-target, or an error might have been found in the rules governing queries.

It is possible, however, to have very coarse-grained controls on the data held by analysts, controls that implement the existing U.S. government information classification system. Indeed, the IC has supported research on such controls since the 1970s, under the rubric of "multi-level security." More recent academic work calls it "information flow control." It is quite well understood in theory, and several systems have been built that enforce the rules for handling classified data. Unfortunately, attempts to use these systems in practice have been unsuccessful, and almost none are deployed. Information flow control cannot do the kind of query-specific control that is described in this chapter; instead, it tends to push computed outputs to the highest level of classification, which is not useful in practice. However, it is the best technique known at present.

5.3 MANUAL CONTROLS

To ensure compliance with the rules laid down by the legal authorities under which it operates, NSA has a system of internal auditing and oversight, combining automated and redundant human components. The system covers all parts of the foreign intelligence collection system: storage, querying, analysis, and dissemination.

Technology is used to some extent to implement the legal framework for foreign intelligence information, in the form of access controls, secure databases, and an automatically generated audit trail.

There is also extensive human review of all actions, both internal and external. NSA's compliance program is supported by more than 300 personnel across the agency, which includes the Office of the Director of Compliance (established in 2009).[4] Internal oversight is provided by the NSA's Office of Inspector General and the Office of General Counsel. NSA also has a Civil Liberties and Privacy Office, first established in January 2014 shortly after the President's speech on signals intelligence.[5] The other major staff organizations that have responsibility for some facets of civil liberties and privacy responsibilities are the Office of the Director of Compliance, the Authorities Integration Group, and the Associate Director for Policy and Records. The Office of the Director of National Intelligence (ODNI) has its own Civil Liberties and Privacy Office,[6] and the ODNI Office of General Counsel and the Department of Defense Office of General Council have responsibility for oversight as well.

Continuing external oversight is provided by the Department of Justice, congressional oversight committees, and the Foreign Intelligence Surveillance Act (FISA) court. The Privacy and Civil Liberties Oversight Board (PCLOB)[7] and the Intelligence Oversight Board of the President's Intelligence Advisory Board have also examined NSA operations from

---

[4] Office of the Director of National Intelligence, *DNI Clapper Declassifies Intelligence Community Documents Regarding Collection Under Section 501 of the Foreign Intelligence Surveillance Act (FISA)*, 2014, http://www.dni.gov/index.php/newsroom/press-releases/191-press-releases-2013/927-draft-document.

[5] National Security Agency Central Security Service, *NSA Announces New Civil Liberties and Privacy Officer*, January 29, 2014, https://www.nsa.gov/public_info/press_room/2014/civil_liberties_privacy_officer.shtml.

[6] See Office of the Director of National Intelligence, Civil Liberties and Privacy Office, "Who We Are," http://www.dni.gov/index.php/about/organization/civil-liberties-privacy-office-who-we-are, accessed January 16, 2015.

[7] Privacy and Civil Liberties Oversight Board, *Report on the Telephone Records Program Conducted under Section 215 of the USA PATRIOT Act and on the Operations of the Foreign Intelligence Surveillance Court*, January 23, 2014, available at http://www.pclob.gov/library/215-Report_on_the_Telephone_Records_Program.pdf; and Privacy and Civil Liberties Oversight Board, *Report on the Surveillance Program Operated Pursuant to Section 702 of the Foreign Intelligence Surveillance Act*, July 2, 2014, http://www.pclob.gov/library/702-Report.pdf.

CONTROLLING USAGE OF COLLECTED DATA   65

a privacy and civil liberties standpoint. None has found any deliberate attempts to circumvent or defeat these procedures, although there have been documented incidents of error.

Purely automatic control of usage would mean that the rules would be enforced automatically using published mechanisms. Then people outside the IC concerned about privacy and civil liberties would not have to trust that the IC has adequate procedures and follows them, which many of them are reluctant to do. Such purity is not possible, however; it is thus necessary to independently audit the IC's procedures to some extent. The impractical alternative is to make every step on the path from raw data to query results secure from any possible tampering; this would be a rigid and unworkable system. Some manual controls are necessary to ensure that the automatic controls are actually imposed and that they are configured according to the rules, and to decide cases that are too complex to be automated.

Thus, the goal of reassuring the public by the exclusive use of transparent automatic controls is elusive. Those who do not trust the power of government, both its elected officials and the IC, will argue that its technical expertise could be misused to override automatic controls, and no amount of manual or automatic oversight is likely to reassure them. In short, perfect controls are impossible. The goal should be to balance controls against practicality, recognizing that some amount of risk, tempered by trust in those who manage the system, will always remain.

5.4 AUTOMATIC CONTROLS

A technical system for controlling usage of bulk data has three parts: *isolating* the bulk data so that it can only be accessed in specific ways, *restricting* the queries that can be made against it, and *auditing* the queries that have been done. All three parts are equally important, although isolation is most fully developed and hence has the fullest description, and auditing is the least developed. This chapter gives brief descriptions of each of these parts. It emphasizes the architecture of the possible systems, giving only a sketch of the technical details; consult the references for the full story.

Note that any technical mechanism must be tested under realistic conditions to establish confidence that it actually works. This is especially important for mechanisms that are intended to handle rare events, like the ones described here. The only practical way to do this is to deliberately inject disallowed queries into the running system and verify that they are detected and handled correctly.

Some of the methods described here are in widespread use commercially, and perhaps within NSA. Others have been demonstrated in the laboratory at a moderate scale. Except for full homomorphic encryption, it should be possible to deploy any of them at scale within the IC in the next 5 years. However, the committee is not recommending deployment of any of them. Whether more powerful automatic controls should be deployed is a policy question. The answer depends both on the cost to the IC in dollars and in reduced capability, and on how important it is to have better controls. The committee notes that, in some cases, better technology could reduce the cost of existing controls.

5.4.1 Isolation

Isolating bulk data is one technical method for controlling usage. Isolation also makes it easy to log all the queries against it and their results, which is essential for the auditing discussed in Section 5.4.3. Figure 5.2 shows the elements of this method, which is closely related to the standard access control method used in cybersecurity. The bulk data is cut off from the outside world by an isolation boundary. The only way to cross this boundary is to submit a query to the guard, which is responsible for enforcing the policy that says what queries and results are allowed. The guard logs all queries and results for later auditing, and the audit log itself is isolated to protect it from tampering. The isolated domain is hosted by some mechanism that guarantees the isolation; in some sense it runs on the host, and its security therefore depends on the host operating correctly. There are many such mechanisms: airgaps, operating systems, etc.; some of them are discussed below. In all cases, the host is implementing the isolation; if it does not work correctly, the bulk data will not be pro-

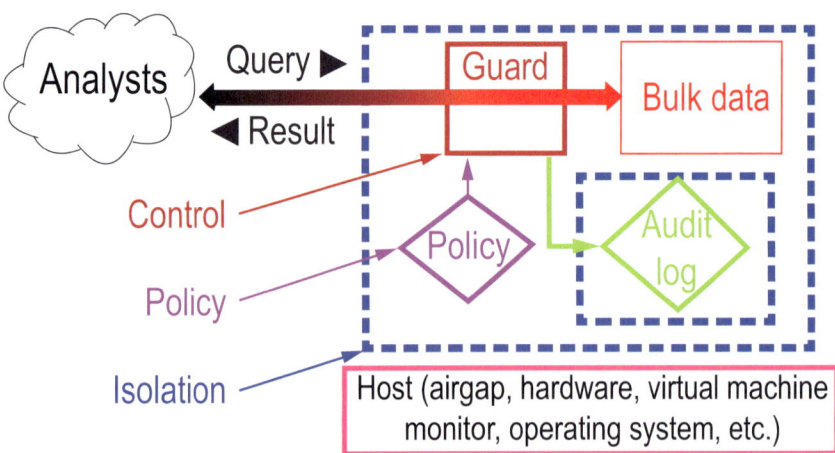

FIGURE 5.2 Isolating bulk data.

tected. For example, if an operating system is corrupted by malware, it will not properly isolate the application processes that it hosts.

Note that isolation depends on the guard as well as the host. If the guard lets through inputs that it should have blocked, the bulk data will not be properly protected. See Section 5.4.1.3.

The critical points in this architecture are the bulk data processing itself, the guard, and the host that implements the isolation boundary. These constitute the trusted computing base (TCB), the parts of the system that must work correctly for the system to be trustworthy. The smaller and simpler they are, the more likely they are to be correct, making it easier for manual review and automated tools to check for mistakes.

Bulk data processing is trusted to correctly implement a query, rather than return something else that might violate the policy. Again, if the TCB is simple, it is easier to understand and more likely to work. There are ways to implement the system so that the most complicated parts are kept outside the TCB; they are described below.

The guard is critical, no matter how the isolation is done; it is trusted to correctly enforce the policy and block malformed inputs. The latter is difficult if the inputs are too complicated for the guard to fully understand. Simple policies and simple inputs make it much more likely that the guard will work correctly. With simple inputs, the guard can concentrate on the job of making sure that all the queries it passes are allowed by the policy. With complicated inputs, it is much easier to hide some piece of malware that is not really a query at all. A familiar example of this is executable malware included in an email message. A policy that says to only accept email from friends is not enough, because friends might be infected themselves. The guard needs to block all executable content that has not been properly vetted.

Traditionally in computer security, the guard implements an *access control* policy that, as shown in Figure 5.2, would specify which analysts are allowed to access which items of bulk data, by attaching to each data item some description of the analysts authorized to access it. NSA has reported that its analysts use some variant of this scheme within a private cloud.[8] Although it is a useful line of defense, this mechanism cannot express more complex policies such as, "Report all contacts that are one hop away from this target and were in Afghanistan during the communication."

---

[8] Dirk A.D. Smith, "Exclusive: Inside the NSA's private cloud," *Network World*, September 29, 2014, http://www.networkworld.com/article/2687084/security0/exclusive-inside-the-nsa-s-private-cloud.html.

5.4.1.1 Federation of Nongovernment Parties

A partly technical approach to isolating bulk data is *federation*: when possible, leaving the data in the hands of multiple parties that are not part of the government; these might be the parties that acquire the data in the first place, such as telephone companies or other communication service providers, or they might be independent third parties. Querying the database then requires querying all the relevant parties and combining the results, as shown in Figure 5.3. In general, it cannot be done in parallel, and it might require repeated queries to the same party. For example, when tracing out a chain of communication in which different links come from different providers, each link may require a separate query. Furthermore, some kinds of preprocessing of the data may be much less effective, for example, working out all the tightly knit cliques of people who communicate with each other a lot, so that it is possible to quickly find all the cliques that an individual belongs to.

Federation has clear advantages for safeguarding privacy and enforcing policies:

• The federated parties are separate from the intelligence agency and may have no incentive to break the rules, which would help reassure those who are concerned that NSA may have incentives to break the rules.[9]

• One party's misbehavior exposes only some of the collected data.

Federation also has clear drawbacks for intelligence:

• The federated parties may have no incentive to cooperate, even if paid; indeed, their customers may object to such cooperation.

• If a federated party is compelled to collect data it otherwise would not, it may introduce privacy risks.

• If forced to cooperate, a federated party may be slow and clumsy, because it is being asked to do things that are not part of its normal business. This may make it difficult to get good results. Note that federation is much more difficult to implement than the cloning of call detail records that is the current practice under FISA Section 215.

• Federated queries may be much slower and less reliable than centralized ones, both because communicating across organizational boundaries is slow and because database-wide optimizations may be impossible, as described above.

---

[9] This is the process used in wiretap investigations authorized by the 1986 Pen Register Act (Title III of the Electronic Communications Privacy Act).

CONTROLLING USAGE OF COLLECTED DATA   69

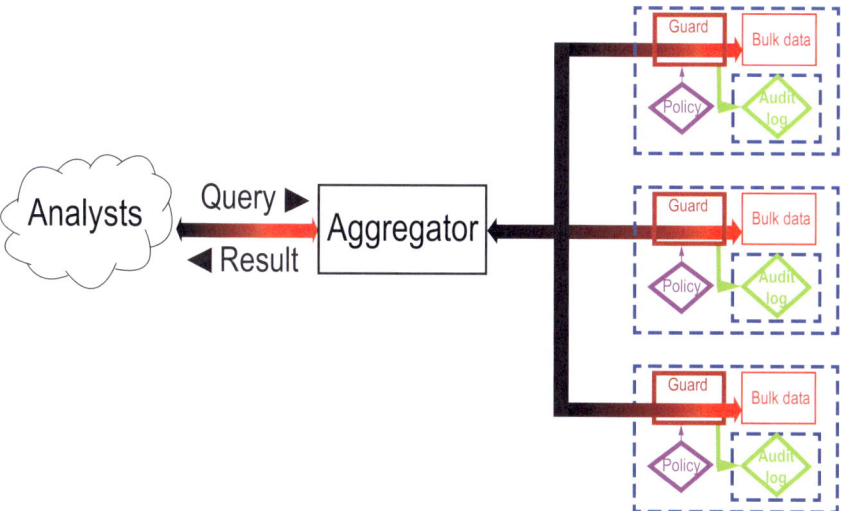

FIGURE 5.3 Federation of non-government parties.

In addition, federation makes the collected data both more and less secure. It is more secure because breaking into one party exposes only some of the data. It is less secure because some of the federated data is exposed if the adversary breaks into any one of the parties.

5.4.1.2 Hosts and Isolation Boundaries

The choice of hosts depends on two things: the acceptable cost of isolation (both capital cost and reduced performance) and the threats it must defend against. More severe threats incur a higher cost, of course, but depend less on manual controls. There are many possible implementations of isolation boundaries with different security strengths and weaknesses and different costs. Here are a few examples:

• *Airgap.* The most secure and most expensive isolation boundary is an airgap: separate physical machines, or networks of physical machines, inside and outside the isolation boundary that is breached only by a carefully controlled network connection. The airgap is costly, because there are two networks of machines to buy and maintain, and the connection between them may be slow. The IC has traditionally used airgaps to isolate classified from unclassified systems; perhaps they are also using airgaps to isolate collected data from analysts. Note that although an airgap is a very good isolation boundary, the isolation also depends on the guard that is supposed to check all inputs, as discussed below.[10]

- *Hypervisor.* A cheaper host is a hypervisor that implements separate virtual machines instead of separate physical ones. Currently, the hypervisor is part of the TCB, and, unfortunately, commercial hypervisors are rather complicated because their main selling point is performance rather than security. But this is cheaper than the airgap because there is only one physical machine, and the bandwidth of communication between the virtual machines can be close to the full memory bandwidth. There are many variations on the hypervisor idea, with different costs and security considerations.[11]
- *Enclaves.* In between separate physical machines and separate virtual machines is a fairly new way of doing isolation, called an *enclave* in the implementation, developed by Intel. This is like a virtual machine, but its isolation is provided directly by the central processing unit (CPU). Because this mechanism is tightly integrated into the CPU and the memory system, it can provide good performance much more simply than a hypervisor.[12]

Language virtual machines. Programs written in languages intended for web pages, such as Java and JavaScript, are usually executed inside isolation boundaries with names like Java Virtual Machine. In this case, the main purpose of the isolation is to protect the rest of the system from the untrusted web program rather than the other way around.

5.4.1.3 The Guard

If the isolation mechanism is sound, the guard is the main weak point; if the guard makes the wrong decisions about what to allow through, the system inside the isolation boundary can be completely compromised, and this has happened many times in practice with every kind of isolation boundary, including airgaps. For example, executable malware included in an email message can infect an isolated system. The same thing can happen with a USB flash drive, which can contain malware that is executed automatically. The guard needs to block all executable content that has not been properly vetted. As with every aspect of security, the

---

[10] Although a good example of an isolation technique, technology alternatives listed in this subsection can be engineered to provide adequate isolation for this application.

[11] M. Pearce, S. Zeadally, and R. Hunt, Virtualization: Issues, security threats, and solutions, *ACM Computing Surveys* 45(2), Article No. 17, 2013.

[12] F. McKeen, I. Alexandrovich, A. Berenzon, C. Rozas, H. Shafi, V. Shanbhogue, and U. Savagaonkar, Innovative instructions and software model for isolated execution, in *Proceedings of the Second International Workshop on Hardware and Architectural Support for Security and Privacy*, Association of Computing Machinery, New York, N.Y., 2013.

only practical approach today is to keep both the specification of what the guard has to do and the code that does it as simple as possible.

If each item of bulk data is tagged with access control information that specifies which analysts are allowed to see it, the job of the guard is easier. The Apache Accumulo open-source database, for example, has this feature; it was originally developed by NSA, which transferred it to Apache, an organization that develops open-source software for the Internet. This kind of tagging is the standard way of doing access control in computer security; it is helpful for controlling usage of collected data, but not sufficient for enforcing a rule such as "trace contacts for at most two hops," which restricts the algorithm that processes the data rather than access to the data itself.

5.4.1.4 Bulk Data Processing

In general, there are a lot of bulk data, so that simply storing the bits securely and reliably is complex, and the data are processed by a general-purpose database system that is even more complex, usually tens of millions of lines of code. Much of this code might not be needed for a particular application, but it is likely to be impractical to separate the parts that are needed from the rest. Thus, it is highly desirable to keep as much of this storage and processing out of the TCB, which should be small and simple.

There has been a lot of work on isolation to protect a cloud client from its cloud service provider, because there is a big market for cloud computing, and many customers care about the security of their data and do not want to trust the service provider. This is the most important application for the enclaves described above. Figure 5.4 shows another way for a client to store and process data in the cloud without trusting the cloud provider. The idea is to do everything in the cloud in encrypted form, so that the result appears in encrypted form as well. Only the client holds the key, so only the client can see anything about the data or the result except its size, and perhaps something about the shape of the query. This gives no guarantee that the result is correct or that it reads only the data actually needed for the query, but it does guarantee that only the client sees any data, and since the client is entitled to see all the data, the client's secrecy is maintained. It is not obvious how to actually implement this scheme, but in some cases it is possible, and ways to do it are explained below.

Unfortunately, although the architecture shown in Figure 5.4 serves the needs of the cloud client, it is not enough for automatic control of access to bulk data. Unlike the cloud client, the analyst is not entitled to see all of the data. How can the guard enforce the policy about what the analyst is allowed to see? This takes a *proof*, or perhaps some convincing

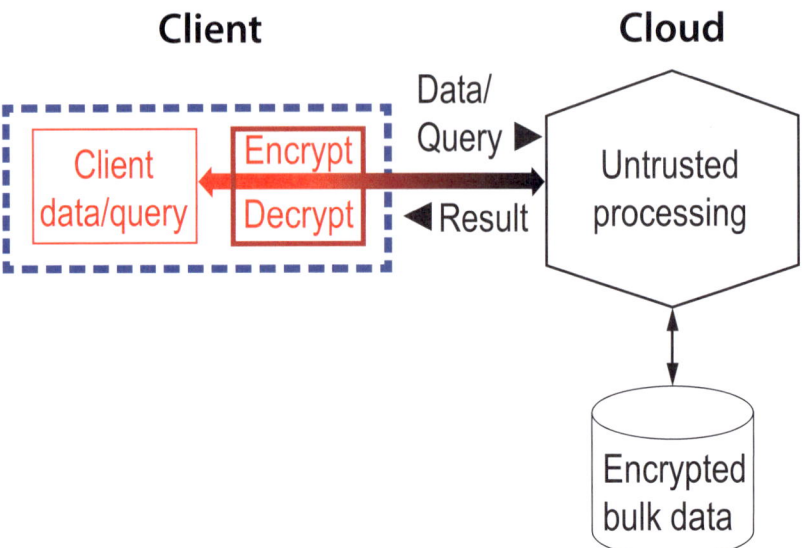

FIGURE 5.4 Smaller trusted computing base by processing encrypted data for a client.

evidence, provided by the untrusted cloud side of the picture, that the result is correct or at least that it does not reveal any *more* data than the query demands. Figure 5.5 illustrates this approach; the parts that are unchanged from Figure 5.2 are dimmed.

What would such a proof look like? That depends on the query. For example, if the query is "Return all the endpoints of communications with this target," a proof would be a list of all the database entries that yielded the result; recall that these are all encrypted, so the untrusted side cannot make them up. If the target is X, and each database entry represents a call detail record with a triple <from, to, time>, verifying the proof means checking that every result endpoint Y is in an entry <X, Y, time> or <Y, X, time>. Note that this does not prove that the result is correct, but it does prove that no extra information is disclosed. For another example, see the next section.

5.4.1.5 Encrypted Data at Rest

The simplest example of the idea in Figure 5.5 uses the untrusted side only to store data, not to do any computing on it, as shown in Figure 5.6 (where the unchanging left side of the figure has been cut off). This means that each data block is encrypted before being handed over to untrusted storage by the collection system and decrypted when it is read back. Each

CONTROLLING USAGE OF COLLECTED DATA 73

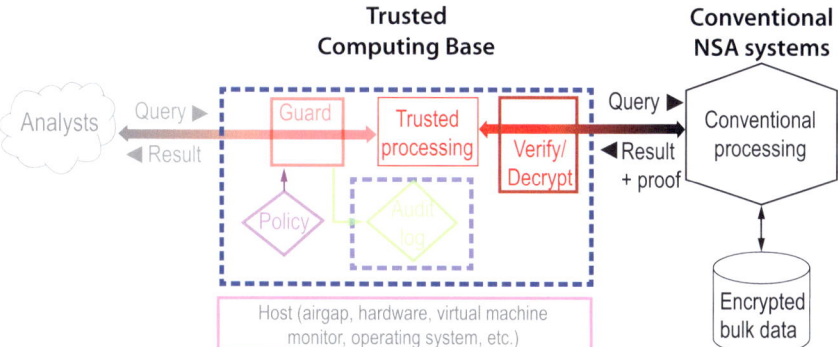

FIGURE 5.5 Smaller trusted computing base by verifying untrusted processing of a query.

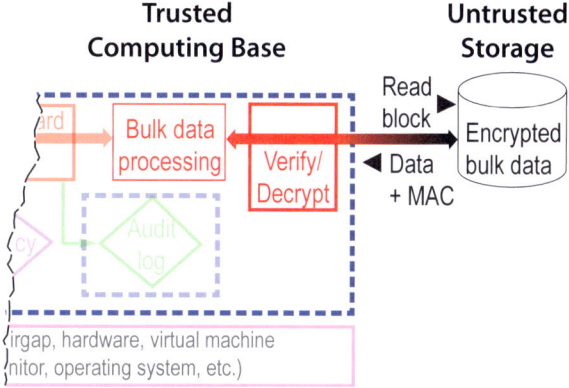

FIGURE 5.6 Smaller trusted computing base by encrypting bulk data at rest.

data block also has a message authentication code (MAC), a well-known cryptographic mechanism that the trusted verifier can use to check that the encrypted data it reads back is indeed the same data that it wrote earlier. The MAC serves as the proof that the untrusted storage is returning the correct result of the read. This scheme has been widely implemented, and it removes the storage hardware from the TCB. It also removes a lot of software, because reliably and efficiently storing large amounts of data is complex, and it takes millions of lines of code to deal with this complexity.[13]

---

[13] Ken Beer and Ryan Holland, "Securing Data at Rest with Encryption," Amazon Web Services white paper, November 2013, http://media.amazonwebservices.com/AWS_Securing_Data_at_Rest_with_Encryption.pdf.

5.4.1.6 Simulating Homomorphic Cryptography

A fancy form of cryptography called homomorphic encryption makes it possible to do *all* the processing on encrypted data, producing an encrypted result without exposing any data in the clear.[14] It is quite surprising that this works at all, but it turns out there are theorems showing that any computation can be done in this way. Doing the untrusted processing with homomorphic encryption provides a complete implementation of Figure 5.4. Unfortunately, the best known ways of doing it, in general, are at least a million times too slow to be practical.

For queries that only need to test whether two values are equal, simple deterministic encryption is sufficient. Perfect encryption would reveal nothing at all about the data values except their approximate size. Deterministic encryption reveals only which values are equal. For queries that need to test whether one value is less than another, order preserving encryption is sufficient. It is more expensive and of course reveals the relative order of the values. Many practical queries fall into one of these categories, and it is not too hard to modify an existing database system to make these queries work entirely on encrypted data.[15] Work using an encrypted search may yield useful results in the future; see Section 6.3.1.

The idea behind homomorphic encryption is that any basic computation on encrypted data, such as adding two numbers, comparing two strings for equality, or sorting a list of items, can be done (slowly) directly on the encrypted data. A practical alternative is to add a small component to the TCB that decrypts the data, does the operation, and encrypts the result, as shown in Figure 5.7; compare this with Figure 5.5. Because most of the basic operations are simple, this component can be small and simple. Indeed, for many applications, it can be simple enough to be implemented in special purpose hardware using field programmable gate arrays, which is both fast and difficult to infect with malware.[16]

---

[14] Craig Gentry, Computing arbitrary functions of encrypted data, *Communications of the ACM* 53(3):97-105, 2010.

[15] R.A. Popa, C.M.S. Redfield, N. Zeldovich, and H. Balakrishnan, CryptDB: Protecting confidentiality with encrypted query processing, in *Proceedings of the Twenty-Third ACM Symposium on Operating Systems Principles*, Association of Computing Machinery, New York, N.Y., 2011.

[16] A.A. Arasu, S. Blanas, K. Eguro, M. Joglekar, R. Kaushik, D. Kossmann, R. Ramamurthy, P. Upadhyaya, and R. Venkatesan, Secure database-as-a-service with Cipherbase, in *Proceedings of the 2013 ACM SIGMOD International Conference on Management of Data*, Association of Computing Machinery, New York, N.Y., 2013.

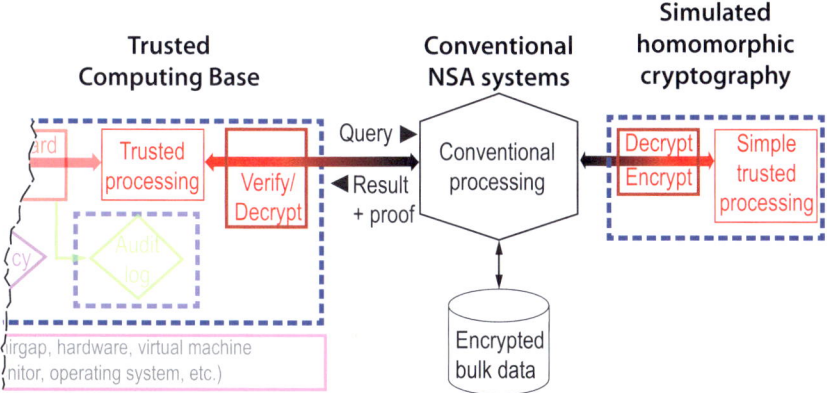

FIGURE 5.7 Smaller trusted computing base by simulating homomorphic cryptography.

5.4.2 Restricting Queries Automatically

Restricting queries automatically is another way to control usage. The goal is to do this well enough that software can decide which queries are allowed by the policy, or at least drastically reduce the number of queries that require human approval. The conventional access control discussed above is one way to do this, but there are many policies that it cannot express. Automated restriction is certainly feasible for limited classes of queries such as, "Find all the phone numbers that have connected in the last month to this list of numbers belonging to a known target." Indeed, NSA already has pre-approved queries, but their scope can probably be extended significantly. The more mechanized the process, the better. Sometimes the software will refer to a human for a decision, but an automated decision is cheaper, faster, potentially more transparent, and less burdensome to analysts. The ideal is that the analyst only sees information about targets, so that there is no intrusion on the privacy of people who are not targets.

Chapter 6 discusses some of the major opportunities for advances here.

5.4.3 Audit/Oversight Automation

Auditing usage of bulk data is essential to enforce privacy protections. The first step is to ensure that every query is permanently recorded in a log. Isolation provides confidence that every query is permanently logged. Then the log must be reviewed for compliance with the rules.

Doing this manually is feasible, and is, indeed, NSA's current practice. Although it is thorough, it is expensive and not transparent—outsiders must rely on the agency's assurance that it is being done properly, because the queries are usually highly classified. Automation of auditing, a direction NSA is pursuing, could both streamline audits and provide assurance to outside inspectors, who can then examine the auditing technology.

The resulting ability to inspect the privacy-protecting mechanisms of the SIGINT process on an unclassified basis may help allay privacy and civil liberty concerns. The inspection would focus on the automation software and the usage rules it enforces, rather than on the data, which must remain classified.

Greater automation of auditing is an area that has been greatly neglected by government, industry, and academia; for example, operating systems write voluminous logs of security-relevant events, but they are seldom looked at, and when they are, a great deal of manual effort is required. Chapter 6 discusses some possible improvements.

5.5 CONCLUSION

This chapter has reviewed a variety of feasible mechanisms, both manual and automatic, for controlling the way that collected data is used. Some of these are deployed in the IC. Others may be deployed, but the committee was not told about them in briefings. All of these mechanisms are feasible to deploy within the next 5 years. Opportunities to introduce enhancements to such capabilities are expected to arise as the information technology systems used for collection and analysis are refreshed and modernized.

Automation of usage controls may simultaneously allow a more nuanced set of usage rules, facilitate compliance auditing, and reduce the burden of controls on analysts. Similarly, there are opportunities to automate the various audit mechanisms to verify that rules are followed. These techniques may permit more of the use controls and audit mechanisms to be explained clearly to the public. It may be possible to express a large fraction of the rules required by law and policy in a machine-processable form that can be rapidly and consistently applied during collection, analysis, and dissemination.

Conclusion 2. Automatic controls on the usage of data collected in bulk can help to enforce privacy protections.

Conclusion 2.1. It will be easier to automate controls if the rules governing collection and use are technology-neutral (i.e., not tied to specific, rapidly changing information and communications tech-

nologies or historical artifacts of particular technologies) and if they are based on a consistent set of definitions.

Conclusion 2.2. Automated controls can provide new opportunities to make the controls more transparent by giving the public and oversight bodies the opportunity to inspect the software artifacts that describe and implement the controls. Increased transparency can give people outside the IC more confidence that the controls are appropriate, although the need for secrecy about some of the details makes complete confidence unlikely.

Whether any given method should actually be deployed is a policy question that requires determining whether increased effectiveness and apparent transparency is worth the cost in equipment, labor, and potential interference with the intelligence mission. In any case, some automatic methods might be able to replace existing manual ones at a lower cost.

6

Looking to the Future

The future of signals intelligence (SIGINT) may look very different from what we see today. Details of communications technologies are changing rapidly and are likely to continue to change. Encryption increasingly protects communications data both in transit and at rest. Private-sector business records may become fewer in number and less useful for intelligence purposes. On the other hand, more powerful computation can analyze raw data, such as speech and images, to extract useful intelligence information in real time. More data will certainly be available, driven by commercial, data-driven marketing as well as the spread of networked sensors of many sorts. Research to develop algorithms to proceed "from data to knowledge" may well benefit intelligence analysis.

The public policy landscape may also change. Public concerns with privacy, driven by the explosion of data as well as disclosures and misadventures in both public and private sectors, may lead to new legal frameworks. A shift from controlling collection to controlling usage is being discussed in policy circles. And, of course, the publicly acceptable trade-off between privacy and security might change immediately if the nation is attacked, if severe national security threats emerge, or if international security is further destabilized.

6.1 THE FUTURE OF SIGNALS INTELLIGENCE

There are a number of trends, already under way today, that may have a deep impact on SIGINT. The net effect of these trends on SIGINT in the future is not clear today.

6.1.1 More Data, Data Types, and Sensors; More Computing and Storage

Declining costs of all elements of digital infrastructure continue to spur technology's pervasive spread. Not long ago, "cloud computing," the use of giant computer centers to assign, as needed, dozens to thousands of computers to a task—was new. Now we are experiencing the effects of "big data," exploiting large amounts of data collected for business or scientific purposes to pursue new business opportunities or uncover new science. Just beginning is an "Internet of things," deploying sensors of new types in many new places to control or optimize roadways, buses, trains, production lines, crop management, and countless other activities. And smart phones increasingly sense things of interest, notably location today, but also audio and video. While not all of this new data, and the communications that carry it, is likely to have intelligence value, some will surely offer new intelligence opportunities.

Increasingly, algorithms can digest raw signal data into much more useful forms. License plates can be located in images taken from roadways, and the license numbers recognized. Faces can be isolated in images captured by surveillance cameras, and databases of images can be queried to identify people.[1] Audio signals of speech can be converted to text with enough reliability for dictation, making it easy to spot words of intelligence interest in communications. These algorithms all have a form that make it technically easy to scale up the processing to handle many inputs: you can assign each of 50 computers to analyze each of 50 license plate images, or you can deploy the same 50 computers to recognize speech. Flexibly adapting and scaling these computations is easy.

In SIGINT applications, these algorithms can be applied either at the time of collection or later, on demand, for analyzing selected data. Today, NSA says it cannot collect any sizeable fraction of all global communications data, and it may likewise be that despite declining computing costs, NSA will not be able to automatically analyze more than a tiny bit. However, in many cases, the operators of the sensors will apply the algorithms to meet business needs, such as identifying license plates to bill parking charges. In these cases, the analyzed data may be available to NSA in the form of business records.

6.1.2 Business Records

Business records can be very valuable for intelligence, and the proliferation of information technology (IT) in businesses of all sorts means

---

[1] Closed-circuit TV surveillance, a form of bulk collection, has been practiced for years with relatively little complaint, despite its privacy invasion.

that many more details of everyday life are recorded in this way. However, businesses that wish to minimize surveillance of their customers can arrange to reduce or eliminate the intelligence value of their records. For example, if a telephone company bills a flat monthly rate, it need not keep a record of each call, so no call data records would be available for intelligence purposes.[2] Communications providers today are acutely aware of their customer's concerns about surveillance,[3] a fact that gives providers an additional incentive to refrain from keeping records that might be used against them.

Services that hold data for customers may find ways to encrypt the data with a key known only to the customer so as to evade surveillance. This technique could be used by email providers and social-networking services, among others. Some businesses are being established with exactly this objective. But today, the ability to examine customer data and use it for marketing purposes is an essential part of the hosting company's business model, so customers are unlikely to have email that is both free and surveillance-proof.

Attempts to evade surveillance are unlikely to slow the big data trend. Businesses collect huge amounts of data not associated with individuals, which may not cause privacy concerns, and are sure to collect still more. Some of this data has a large public benefit, such as for weather prediction, crop management, or public health monitoring. Businesses may implement different levels of protection for different business records, so that customer-sensitive data is not comingled with data that has benign uses, both public and private.

6.1.3 Encryption

One of the most imminent threats to SIGINT collection is the increasing use of strong encryption for signals in transmission. Increasingly, website servers are routinely encrypting traffic to and from the browser clients. To a lesser extent, data at rest is being encrypted. The cybersecurity vulnerabilities of the endpoints (browser, server) are becoming much greater than the vulnerability of the communications between them, a point suggesting that access may still be possible (although more difficult), even when transmission links are encrypted.

---

[2] Other business records of such a company, however, linking customer name, address, and telephone number, might still be very valuable for intelligence purposes.

[3] See, for example, Vodafone Group, "Law Enforcement Disclosure Report," 2014, http://www.vodafone.com/content/sustainabilityreport/2014/index/operating_responsibly/privacy_and_security/law_enforcement.html, accessed January 16, 2015.

6.1.4 Services That Evade Surveillance

Although today's common Internet services, such as VoIP (Voice over Internet Protocol) are not specifically designed to make surveillance difficult, they can be redesigned to evade surveillance. An important idea in many cases is "peer-to-peer" communications, which establishes an encrypted channel between two communicators without needing a third party to set up the communication. This technique means there is no third-party business that might hold business records or other data that could identify the communicators. It can be a bit tricky to design protocols that eliminate the third party, which often serves as a "directory" for a calling party to find the called party. And, of course, it is hard on the third-party business, which is trying to make money when callers communicate.

6.1.5 SIGINT Must Adapt

An unsurprising conclusion from the preceding subsections is that SIGINT techniques and operations will need to evolve as dynamically as the signals environment they monitor. As new protocols and businesses arise, collection methods and software must evolve. Adapting to traffic volume of different types is also essential, but it can be partially addressed by using techniques similar to those used in cloud computing.

Policy, law, and regulations will need to keep up with future SIGINT sources, which may evolve in dynamic and even surprising ways. Today, the laws governing collection of SIGINT are largely derived from legislation that applies to rotary dial telephones. Although policy and regulations have adapted to modern technologies, their pace of change does not match that of technology.

6.2 EVOLUTION OF PRIVACY PROTECTIONS

A striking change in the past few decades is the extent to which the private sector collects personal information. This trend had its origins in the 1960s with the rise of credit bureaus and has resulted in a cascade of law and regulation. In 1998, the Federal Trade Commission published a list of five core principles: (1) notice (give the consumer notice of data collection), (2) choice (give the consumer choice about whether the private data will be collected), (3) access (give the consumer the ability to access data about him- or herself), (4) integrity/security (the data collector must work to make sure data is correct and must give the consumer the right of redress if it is not), and (5) enforcement/redress. Today, these principles are known by the shortened phrase "Notice and Consent."

The notice and consent framework is showing signs of stress. A recent President's Council of Advisors on Science and Technology (PCAST) report ridiculed the turgid privacy terms that the public is typically asked to accept today: "Only in some fantasy world do users actually read these notices and understand their implications before clicking to indicate their consent."[4] Moreover, "consent" may imply that a person is volunteering personal data, which will mean it is afforded weaker Constitutional protection.

An alternative, which is starting to be discussed in policy circles, is to control *use* rather than collection of data. One variant calls for tagging all data with its origin and asking permission of its provider before using it. The data can be encrypted, so that only the provider's grant of permission will reveal the actual data. Protecting use of data is not new; digital rights management (DRM) schemes encrypt songs and videos and only decrypt and play them when the player is given the key. Changing to protecting use of private data would be a major effort, requiring changes in laws and enforcement and, of course, a lot of software.[5]

Protecting use rather than limiting the collection of sensitive data would be consistent with maintaining the bulk collection of SIGINT. Perhaps if the public comes to embrace the philosophy and practice of usage controls for sensitive personal data, such as health and financial data, and comes to trust private sector IT implementations of the protection procedures, controlled-use approaches to intelligence information can find greater favor.

6.3 RESEARCH AND DEVELOPMENT

This section contains a collection of topics that came up during the committee's deliberations that are potentially useful to the IC. None of these topics directly addresses ways to replace bulk collection with targeted collection. Because the main focus of this report was not to determine the full set of research areas to explore, this list is not meant to be complete.

Research is under way on all of the topics mentioned in this section. In many cases, NSA already implements some of the capabilities (e.g., certain kinds of query checking). The IC has research efforts under way in many of these areas as well. Of particular note is the Intelligence Advanced Research Projects Activity's (IARPA's) Security and Privacy

---

[4] President's Advisory Committee on Science and Technology (PCAST), *Big Data: A Technological Perspective*, May 1, 2014, http://www.whitehouse.gov/sites/default/files/microsites/ostp/PCAST/pcast_big_data_and_privacy_-_may_2014.pdf, p. xi.

[5] Craig Mundie, Privacy pragmatism: Focus on data use, not data collection, *Foreign Affairs*, March/April 2014, http://www.foreignaffairs.com/articles/140741/craig-mundie/privacy-pragmatism.

Assurance Research (SPAR[6]) program, which addressed topics of particular relevance to implementing secure SIGINT systems of the sort described in Chapter 5.

This section does not delve into the many technologies that NSA and other IC organizations use to operate large, complex IT operations. It does not cover network security, operating-system security, physical security of computer systems, authentication of users, or a host of other areas that are part of making SIGINT technologies trustworthy. Research in these and other areas that affect the general state of complex IT will help the IC too.

6.3.1 Technologies for Isolation

The approaches described in Section 5.4 are not in widespread use, but they are not unexplored either. Their successful use will depend on not only choosing a sound architecture, but also on developing a careful implementation: the trustworthiness of key components depends on keeping them simple to avoid mistakes that lead to vulnerabilities. And system-wide properties, such as security, will depend on many details, such as managing cryptographic keys properly, distributing them securely, changing them occasionally, ensuring that no single system administrator can penetrate security, and so on. These are not simple systems to engineer and operate.

Variants of the systems described in Chapter 5 often involve executing separate components on separate computers (often under control of separate organizations) and protecting the communications among the components. Techniques for doing this, usually based on encryption, are the topic of a research area dubbed "secure multi-party computation," which was investigated by the IARPA SPAR program. For example, recent research shows how to protect data and communications in a three-part system: one issues queries, a second authorizes queries, and a third holds data and performs searches specified by authorized queries.[7]

6.3.2 Other Technologies for Protecting Data Privacy

Although the focus of this report is signals intelligence that provides data about individual people and groups, signals intelligence can also

---

[6] Office of the Director of National Intelligence, Intelligence Advanced Research Projects Activity, "Security and Privacy Assurance Research (SPAR)," http://www.iarpa.gov/index.php/research-programs/spar, accessed January 16, 2015.

[7] S. Jarecki, C. Jutla, H. Krawcyzk, M. Rosu, and M. Steiner, Outsourced symmetric private information retrieval, pp. 875-888 in *Proceedings of the 2013 ACM SIGSAC Conference on Computer and Communications Security*, Association of Computing Machinery, New York, N.Y., 2013.

be used to answer questions such as "What is the most common disease mentioned in Internet search requests from Yemen?" In these cases, the question is statistical, and protecting the identities of people cited in the source database can be done using techniques quite different from those prescribed for tracking specific threats. Although it might seem that statistical questions by their nature do not reveal identities, if the query specifies a sufficiently small group, identities can often be inferred using queries to different databases.

Collecting and publishing large data sets ("open data") has spurred work on ways to benefit from the data without revealing personal information. One class of techniques attempts to "anonymize" (or de-identify) the data by transforming it to retain useful information but prohibit identification of individuals. But it turns out that most anonymization schemes are easy to defeat.[8] Effective anonymization remains an open problem.

Differential privacy is an active research area tackling the problem of enabling statistical queries from collections of data while preserving the privacy of individuals.[9] The purpose is to permit useful information to be determined while not exposing data on specific individuals, including individuals not included in the data. This is done by adding probabilistically structured noise (small probabilistic changes to the data) to the responses to the queries. Although statistical databases have value in many domains, the type of queries relevant to this report need to produce information about individual items, so the techniques of differential privacy are not immediately applicable. There is also work on using differential privacy techniques with social networks.[10]

6.3.3 Approving Queries and Their Results Automatically

Automatically restricting or approving a query requires automatically understanding it at a deeper level than syntax; this points to another advantage of automated decision making, namely, that it forces precision about what is being collected, which is useful both for analysts and for privacy. Automated understanding can be either static or dynamic. *Static*

---

[8] See PCAST, *Big Data,* 2014, p. 38. A good view of anonymization and reidentification is in Sections 3 and 4 of "Opinion 05/2014 on Anonymisation Techniques" (European Commission, Article 29 Data Protection Working Party, adopted April 10, 2014, http://ec.europa.eu/justice/data-protection/article-29/documentation/opinion-recommendation/files/2014/wp216_en.pdf).

[9] Cynthia Dwork and Aaron Roth, *The Algorithmic Foundations of Differential Privacy*, Now Publishers, Boston, Mass., 2014.

[10] C. Task and C. Clifton, A guide to differential privacy theory in social network analysis, pp. 411-417 in *Proceedings of the 2012 IEEE/ACM International Conference on Advances in Social Networks Analysis and Mining (ASONAM)*, IEEE Computer Society, Washington, D.C.

understanding tries to infer from a set of axioms whether a particular query or class of queries is allowed by policy, independently of the state of the database being queried. Taking pre-approved queries as axioms is a simple case of this. For example, "If X is an identifier for a reasonable and articulable suspicion (RAS) target, return all the identifiers that communicated with X in the last year." The query is fixed except for some parameters, such as X in this example. It is a human decision to pre-approve it, and no automated reasoning is needed to apply it. A more powerful system could deduce that this query is OK from more general axioms such as, "X is associated with Y if X and Y communicated in the last year" and "Any identifier associated with a RAS target can be disclosed."

Dynamic understanding looks at the *actual* results of a query, rather than considering all possible results, and asks whether policy allows them to be disclosed. A simple example is a kind of minimization: if a query returns a set of identifiers, any identifiers for U.S. persons should be removed from the results. *Tags* on the data that track its provenance or other properties can make dynamic understanding more powerful; the example uses a "U.S. person" tag that is added to database entries for an identifier when it is determined that the identifier refers to a U.S. person.[11] This kind of understanding has been studied extensively in the context of information flow control, where the goal is to keep secrets from being disclosed to uncleared people, even if it is processed by untrusted programs. Decentralized information flow can very flexibly represent both degrees of secrecy and authorities for disclosure.[12] Dynamic systems can also take account of context and history by applying the rules in force at the time a query is made, considering questions such as, Is there an emergency? Is the query part of a pattern known to need more scrutiny? Are the results being combined with other data to deanonymize the results in a way that is contrary to policy?

There are many similarities between static and dynamic understanding and the thriving fields of static and dynamic program understanding, which suggests that there may be rich opportunities here. Not surprisingly, programs written in languages that are designed for automated understanding are much easier to understand. The same thing applies to query languages; indeed, the standard SQL database query language is

---

[11] NSA has developed and donated to the Apache open-source community such a database. Accumulo is a scalable key/value store that allows "access labels" to be attached to each cell that enables low-level query authorization checks (Apache Software Foundation, "Apache Accumulo," https://accumulo.apache.org/, accessed January 16, 2015).

[12] A.C. Myers and B. Liskov, A decentralized model for information flow control, pp. 129-142 in *Proceedings of the 17th ACM Symposium on Operating System Principles (SOSP)*, 1997, Association for Computing Machinery, New York, N.Y.

designed for automatic understanding of queries, and database systems make heavy use of this to optimize their execution.[13] A system that can understand a query can also rewrite it to add access control checks or calls to functions that encrypt and decrypt sensitive fields. For example, the CryptDB and Cipherbase systems do this (see Section 5.4.1.6 on simulating homomorphic encryption).

In most cases, a query is issued by an analyst, and the results are returned to the analyst, but there are also programs, called *analytics* by the IC, that issue queries and process the results themselves. Understanding these analytics programs requires combining an understanding of the queries with an understanding of the program that issues them. However, the issuing program can supplement the query itself with additional information that can be used in making a decision whether to approve the query. In other words, the program generating the query can be expected to do more work to support a decision whether to approve the query than might be practical for a human analyst.

The most likely approach to query approval is to proceed from easy cases to harder ones, reserving for human attention those that cannot be automated.

6.3.4 Audit/Oversight Automation

Auditing access to bulk data is essential for ensuring compliance with the rules. The first step is to ensure that every query is permanently recorded in a log; isolation makes it feasible to do this by technical means. Then the log must be reviewed. Doing this manually is feasible and, indeed, this is NSA's current practice, but it is expensive and not transparent—outsiders must rely on the agency's assurance that it is being done properly, because the queries are usually highly classified.

In analogous fashion, operating systems and networking equipment write voluminous logs of security-relevant events, and review of such logs usually requires a great deal of manual effort.[14] It should be possible to develop much better tools that automatically review the log, highlight

---

[13] S. Chaudhuri, An overview of query optimization in relational systems, pp. 34-43 in *Proceedings of the ACM Symposium on Principles of Database Systems,* 1998, Association for Computing Machinery, New York, N.Y.

[14] See, for example, USENIX, WASL '08, First USENIX Workshop on the Analysis of Systems Logs, "Workshop Sessions," https://www.usenix.org/legacy/event/wasl08/tech/ (last changed January 26, 2009) and, to infer causality, see M. Chow, D. Meisner, J. Flinn, D. Peek, and T.F. Wenisch, "The mystery machine: End-to-end performance analysis of large-scale Internet services, pp. 217-231 in *11th USENIX Syposium on Operating Systems Design and Implementation*, October 2014, https://www.usenix.org/conference/osdi14/technical-sessions/presentation/chow.

suspicious patterns, filter out the great majority of queries that do not raise any issues or that were vetted by automatic query approval, and present the remainder for manual review.

Automating the audit or overview process has much in common with automating query authorization. Because there is a lot of audit data, machine learning can also play a role, although it would probably require introducing a lot of synthetic misbehavior (that is, deliberately introduced misbehavior) to get enough true positives into the training set.

6.3.5 Formal Expression of Laws and Regulations

If it were possible to express the laws, policies, and rules governing SIGINT in a machine-understandable form, it might be possible to generate tools that do automatic approval and oversight for a portion of the queries. One approach would be to develop formal policy languages to represent the precise meanings of policies. These could serve as an intermediate language between the output of lawyers and the technological control of processes and computer programs. The process of formulating them would likely reveal many anomalies, ranging from ambiguities to misinterpretations to inconsistencies. NSA reported that it had looked into deontic description logic for this purpose. To the extent that the field of computational law thrives, its results would be relevant. Projects around this area would seem to be an ideal unclassified research topic, appropriate for an interdisciplinary team of experts in law, policy, and computer science.

Basing automation on formal definitions has another advantage: if the rules must change, the automation will change as a direct consequence. Formal rule expressions will change, due to new laws, policies, and regulations, or in order to adapt to emergencies. Of course, the rule expressions and the process for changing them must be controlled carefully to ensure compliance with the governing documents.

Advances in this area might lead outside organizations to gain confidence that the rules for handling personal data are being followed. If these techniques are not being used today, how might they be applied to reassure overseers that what they see is a full report of what happened? Can zero-knowledge proofs be used in some way to reassure members of the public who wish to monitor operations? Are there general ways of scanning logs and reliably picking out transactions that need to be looked at? Cybersecurity defense tries to do this, but even with specialized logs, it is an incompletely solved problem.

6.3.6 Policy Research

Simpler or more understandable rules are desired, but it is not obvious how to create them, nor how to avoid the processes that produced the existing ones. This sort of research could be done independently of the Intelligence Community (IC), at the risk of irrelevance. Some kind of cooperative research leading to unclassified results would be best.

A seemingly simple, but fundamental, problem is the lack of a common lexicon to define the technology relevant to intelligence as it is controlled by law, regulation, and policy directives. This deficiency came up in many discussions, both inside and outside the IC. The absence of such a consensus on terminology may well explain some of the misunderstandings that exist between the IC, its overseers, and the public. If not addressed, it is likely that this confusion will continue and impede the effective development of a policy and legal framework. More generally, without consensus on terminology, the development of effective regulation of technology will be a continuing problem that also impedes building the necessary public trust in the IC. An interdisciplinary effort to develop common terminology for modern and emerging technology would be worthwhile.

6.3.7 Measuring Effectiveness of Intelligence Techniques and the Value of Data

Policy decisions might be informed by quantifying the benefits of various intelligence-gathering techniques as well as their risks. Anecdotal testimony that cites specific events doubtless understates the value of intelligence and also gives the misleading impression that the value of intelligence is in finding the single piece of evidence that thwarts an attack.[15] More often, small bits of information from different sources contribute to an actionable finding.

The IT systems that produce and record intelligence, especially those used by analysts to bring together the bits and pieces gathered throughout an investigation, can track the provenance of the information. Can investigations, once completed, be mined to estimate the value of different sources of intelligence?

Statistical results and machine learning have a role to play. Statistical techniques allow one to estimate the value of different sources of data. Learning techniques potentially allow one to extract more information (better results) from collected data, or more confidently ignore data that

---

[15] See, for example, Privacy and Civil Liberties Oversight Board, *Report on the Telephone Records Program Conducted under Section 215 of the USA PATRIOT Act and on the Operations of the Foreign Intelligence Surveillance Court*, January 23, 2014, http://www.pclob.gov/SiteAssets/Pages/default/PCLOB-Report-on-the-Telephone-Records-Program.pdf, p. 145 ff.

do not have to be collected. As the number of data sources grows, especially from public information, it may become important to routinely assess the value of these sources. And such analysis would provide, at least in classified form to the IC, an answer to a question that Presidential Policy Directive 28,[16] in effect, asks, "How valuable is bulk collection of domestic telephone metadata?"

6.4 ENGAGEMENT WITH THE RESEARCH COMMUNITY

As the committee did its work, it noted an evolving relationship between NSA and the academic research community on problems such as those addressed in this report. For many years, NSA has formally funded unclassified, basic research in mathematics (algebra, number theory, discrete mathematics, probability, and statistics) in the United States in its Mathematical Sciences Program.[17] According to NSA, this program was initiated in response to a need to support mathematics research in the United States and recognizes the benefits both to academia and NSA accruing through a vigorous relationship with the academic community.

Further developing a similarly vigorous and sustained relationship between NSA and the academic computer science community could have similar benefits. Mechanisms would have to be found to translate classified problems into unclassified ones that researchers could tackle without being subject to security review—doing so would improve the coupling of the research mission with the operational mission. The IC has two mechanisms that help bridge the classification "chasm." IARPA funds research relevant to the IC, some of which targets the future of SIGINT. Many of its research programs are predominantly unclassified, and it is working to develop unclassified "proxies" for research problems of more direct applicability to the IC. The firm In-Q-Tel acts somewhat like a venture fund for innovative technology potentially useful to the IC, supporting commercially viable technologies that might serve IC needs. Both appear to be effective, but their structures and policies are not primarily intended to build long-term and vigorous relationships with academic disciplines. Bridging the chasm would benefit both communities.

Even in a report that was intended to address primarily technical issues, the committee found it necessary to engage with a number of legal and policy issues. This point underscores the fact that it is often

---

[16] The White House, Presidential Policy Directive/PPD-28, "Signals Intelligence Activities," Office of the Press Secretary, January 17, 2014, http://www.whitehouse.gov/sites/default/files/docs/2014sigint_mem_ppd_rel.pdf.

[17] National Security Agency, "Mathematical Sciences Program," last modified August 30, 2013, https://www.nsa.gov/research/math_research/.

important for technical research to be conducted in an interdisciplinary manner cognizant of policy issues. But interdisciplinary work integrating technology, law, and policy remains the exception rather than the rule in academic research institutions. Much more of this type of collaboration is required if law and policy are to effectively manage the challenges being generated by rapidly changing technologies.

6.5 CONCLUSION

The committee has identified a number of technical areas where advances could help the IC address privacy concerns about SIGINT data. None of these topics directly addresses ways to replace bulk collection with targeted collection; rather, they represent alternatives for better targeting collection or better controls on usage after collection. Because determining the full set of research areas to explore was not the main focus of this report, this list is not meant to be complete, and it does not delve into most of the technologies that the IC uses for its IT capabilities. Nor are the topics necessarily new; research may be under way, the IC may already have implemented some of the capabilities, and the IC has research efforts under way in many of these areas as well.

> **Conclusion 3.** Research and development can help in developing software intended to (1) enhance the effectiveness of targeted collection and (2) improve automated usage controls.

> **Conclusion 3.1.** The use of targeted collection can be improved by enriching and streamlining methods for determining and deploying new targets rapidly and using automated processing and/or streamlined approval procedures.

Analytics, such as "big data analytics," may help narrow collection, even if they are not sufficiently precise to identify individual targets. If the government is constrained by privacy concerns to collect less data, it may nevertheless be able to use the power of large private-sector databases, analytics, and machine learning to shape the constraints to collect only data predicted to have high value. New uses by the government of private-sector databases would also raise new privacy and civil liberties questions.

Some of these methods may require a great deal of computing, so that filters should be cascaded to first apply cheap tests, followed by more expensive filters only if earlier filters warrant. For example, if metadata indicates a civilian telephone call to a military unit under surveillance, speech recognition and subsequent semantic analysis might be applied to the voice signal, resulting in an ultimate collection decision. Richer targeting may require enhancing the ability of collection hardware and software to apply complex discriminants to real-time signals feeds.

> **Conclusion 3.2. More powerful automation could improve the precision, robustness, efficiency, and transparency of the controls, while also reducing the burden of controls on analysts.**

Some of the necessary technologies exist today, although they may need further development for use in intelligence applications; others will require research and development work. This approach and others for privacy protection of data held by the private sector can be exploited by the IC. Research could also advance the ability to systematically encode laws, regulations, and policies in a machine-processable form that would directly configure the rule automation.

Appendixes

A

Observations about the Charge to the Committee

The committee makes several clarifying observations about the charge it was given:

- The charge distinguishes explicitly between bulk and targeted collection. In fact, the vast majority of applications interesting to the Intelligence Community demonstrate that bulk collection and targeted collection play complementary roles. Drawing a sharp line between bulk and targeted collection does not accurately reflect how these approaches are used in practice. Furthermore, and as discussed in Section 2.2, bulk and targeted collection exist along a continuum without a bright line to differentiate between them.
- The charge calls for the committee to "use cases" to the extent possible. Although the committee's report does discuss a number of use cases (Chapter 3), and these use cases helped it to understand how bulk collection functions as a part of the analytic process, in the end the committee did not find that these use cases were particularly helpful in identifying or explicating possible alternatives to bulk collection. The committee found it more useful to rely on general principles to reach its conclusions.
- The charge implicitly assumes that technology alternatives could make a contribution to the missions of the intelligence community that is roughly comparable to the contribution that bulk collection makes. As noted above, the committee found that this was not the case—in many cases, bulk collection does in fact make unique contributions to the mis-

sions of the intelligence community that other kinds of collection cannot provide. See Chapter 4.

• The charge asks the committee to develop relevant criteria or metrics for comparing bulk collection to targeted collection. But the committee found that decisions about bulk versus targeted collection—and indeed about all manner of collection decisions—are driven by the concerns of policy makers, which are themselves shaped by their perception of the threat environment. Thus, it is not at all obvious that metrics for comparing bulk collection to targeted collection are particularly relevant in the big picture.

B

Acronyms

CDR	call detail records
CIA	Central Intelligence Agency
FISA	Foreign Intelligence Surveillance Act
FISC	Foreign Intelligence Surveillance Court
http	Hypertext Transport Protocol
IARPA	Intelligence Advanced Research Projects Activity
IC	Intelligence Community
IP	Internet Protocol
ISP	Internet Service Provider
IT	information technology
MAC	Message Authentication Code
NSA	National Security Agency
ODNI	Office of the Director of National Intelligence
PPD	Presidential Policy Directive
RAS	reasonable and articulable suspicion

SIGINT	signals intelligence
SMTP	Simple Mail Transport Protocol
SPAR	Security and Privacy Assurance Research
SQL	Structured Query Language
TCB	trusted computing base
TCP	Transmission Control Protocol
USB	Universal Serial Bus
USSID	U.S. Signals Intelligence Directive
VoIP	Voice over Internet Protocol
WMD	weapons of mass destruction

C

Biographical Information for Committee Members, Consultants, and Staff

COMMITTEE

ROBERT F. SPROULL, *Chair*, is an adjunct professor of computer science at the University of Massachusetts, Amherst. Dr. Sproull retired in 2011 as vice president and director of Oracle Labs, an applied research group that originated at Sun Microsystems (acquired by Oracle in 2010). Before joining Sun in 1990, he was a principal with Sutherland, Sproull, and Associates; an associate professor at Carnegie Mellon University; and a member of the Xerox Palo Alto Research Center. He has served as chair of the National Research Council's (NRC's) Computer Science and Telecommunications Board (CSTB) since 2009. He is also on the Computing Community Consortium (CCC) Council. In June, Dr. Sproull completed a 6-year term on the National Academy of Engineering (NAE) Council. He is a member of the NAE and a fellow of the American Association for the Advancement of Science (AAAS) and the American Academy of Arts and Sciences. Dr. Sproull received his M.S. and Ph.D. in computer science from Stanford University and an A.B. in physics from Harvard College.

FREDERICK R. CHANG is the director of the Darwin Deason Institute for Cyber Security, the Bobby B. Lyle Endowed Centennial Distinguished Chair in Cyber Security, and a professor in the Department of Computer Science and Engineering in Southern Methodist University's (SMU's) Lyle School of Engineering. Dr. Chang is also a senior fellow in the John Goodwin Tower Center for Political Studies in SMU's Dedman College. He has been professor and AT&T Distinguished Chair in Infrastructure

Assurance and Security at the University of Texas, San Antonio, and he was at the University of Texas, Austin, as an associate dean in the College of Natural Sciences and director of the Center for Information Assurance and Security. Dr. Chang is the former director of research at the National Security Agency (NSA). In the private sector, he was most recently the president and chief operating officer of 21CT, Inc., an advanced intelligence analytics solutions company. Earlier, he was with SBC Communications where he held a variety of executive positions, including President, Technology Strategy, SBC Communications; president and CEO, SBC Technology Resources, Inc.; and vice president, network engineering and planning, SBC Advanced Solutions, Inc. Dr. Chang began his professional career at Bell Laboratories. He has been awarded the NSA Director's Distinguished Service Medal. He has served as a member of the Commission on Cyber Security for the 44th Presidency and as a member of the NRC's CSTB. Dr. Chang is also a member of the Texas Cybersecurity, Education, and Economic Development Council. He received his B.A. from the University of California, San Diego, and his M.A. and Ph.D. degrees from the University of Oregon. He has also completed the Program for Senior Executives at the Sloan School of Management at the Massachusetts Institute of Technology (MIT). Dr. Chang is the lead inventor on two U.S. patents (U.S. patent numbers 7272645 and 7633951). He has served as an expert witness for congressional hearings on cybersecurity research and development and the security of healthcare.gov.

WILLIAM DuMOUCHEL is a chief statistical scientist for Oracle Health Sciences, at Oracle Data Sciences. His current research focuses on statistical computing and Bayesian hierarchical models, including applications to meta-analysis and data mining. He is the inventor of the empirical Bayesian data mining algorithm known as Gamma-Poisson Shrinker (GPS) and its successor MGPS, which have been applied to the detection of safety signals in databases of spontaneous adverse drug event reports. These methods are now used within the Food and Drug Administration and industry. From 1996 through 2004, he was a senior member of the data mining research group at AT&T Labs. Prior to 1996, he was chief statistical scientist at BBN Software Products, where he was lead statistical designer of software advisory systems for experimental design and data analysis called RS/Discover and RS/Explore. He has been on the faculties of the University of California, Berkeley, the University of Michigan, MIT, and, most recently, was professor of biostatistics and medical informatics at Columbia University (1994-1996). He has authored approximately 50 papers in peer-reviewed journals and has also been an associate editor of the *Journal of the American Statistical Association*, *Statistics in Medicine*, *Statistics and Computing*, and the *Journal of Computational and Graphical Statistics*. Dr. DuMouchel received a Ph.D. in statistics from Yale University.

MICHAEL KEARNS is a professor in the Computer and Information Science Department at the University of Pennsylvania, where he holds the National Center Chair. Dr. Kearns's research interests include topics in machine learning, algorithmic game theory, social networks, and computational finance. He is the faculty founder and co-director of Penn's Warren Center for Network and Data Sciences and the faculty founder of Penn's Network and Social Systems Engineering program. Dr. Kearns has secondary appointments in the statistics and operations and information management departments of the Wharton School. Until July 2006, he was co-director of Penn's interdisciplinary Institute for Research in Cognitive Science. He has consulted widely for many companies (finance, Internet technologies, etc.) and occasionally serves as an expert witness/consultant on technology-related legal and regulatory cases. During the 1990s, Dr. Kearns worked in basic artificial intelligence (AI) and machine learning research at Bell Labs and AT&T Labs, where he was head of the AI department. He has served on the editorial boards of well-known journals of computer science, machine learning, and game theory. He received his Ph.D. in computer science from Harvard University.

BUTLER LAMPSON is a technical fellow at Microsoft Corporation and an adjunct professor at MIT. He has worked on computer architecture, local area networks, raster printers, page description languages, operating systems, remote procedure call, programming languages and their semantics, programming in the large, fault-tolerant computing, transaction processing, computer security, WYSIWYG editors, and tablet computers. He was one of the designers of the SDS 940 time-sharing system, the Alto personal distributed computing system, the Xerox 9700 laser printer, two-phase commit protocols, the Autonet LAN, the SPKI system for network security, the Microsoft Tablet PC software, the Microsoft Palladium high-assurance stack, and several programming languages. He received the Association for Computing Machinery (ACM) Software Systems Award in 1984 for his work on the Alto, the IEEE Computer Pioneer award in 1996, the von Neumann Medal in 2001, the Turing Award in 1992, and the NAE's Draper Prize in 2004. He is a member of the National Academy of Sciences and the NAE and a fellow of the ACM and the American Academy of Arts and Sciences.

SUSAN LANDAU is professor of cybersecurity policy in the Department of Social Science and Policy Studies at Worcester Polytechnic Institute. Dr. Landau has been a senior staff privacy analyst at Google, a distinguished engineer at Sun Microsystems, a faculty member at the University of Massachusetts, Amherst, and at Wesleyan University. She has held visiting positions at Harvard University, Cornell University, and Yale University, and the Mathematical Sciences Research Institute. Dr. Landau is the author of *Surveillance or Security? The Risks Posed by New Wiretapping Technologies* (2011) and co-author, with Whitfield Diffie, of *Privacy on the Line: The Politics of Wiretapping and Encryption* (1998, rev. ed. 2007). She has written numerous computer science and public policy papers and op-eds on cybersecurity and encryption policy and testified in Congress on the security risks of wiretapping and on cybersecurity activities at the National Institute of Standards and Technology's Information Technology Laboratory. Dr. Landau currently serves on the NRC's CSTB. A 2012 Guggenheim fellow, she was a 2010-2011 fellow at the Radcliffe Institute for Advanced Study, the recipient of the 2008 Women of Vision Social Impact Award, and also a fellow of AAAS and ACM. She received her B.A. from Princeton University, her M.S. from Cornell University, and her Ph.D. from MIT.

MICHAEL E. LEITER is executive vice president for business development, strategy, and mergers and acquisitions at Leidos. Prior to taking on his current role at Leidos, Mr. Leiter was a senior counselor at Palantir Technologies. Before that, he was the director of the National Counterterrorism Center (NCTC). He was sworn in as the Director of NCTC on June 12, 2008, upon his confirmation by the U.S. Senate and after serving as the acting director since November 2007. Before joining NCTC, Mr. Leiter served as the deputy chief of staff for the Office of the Director of National Intelligence (ODNI). In this role, he assisted in the establishment of the ODNI and coordinated all internal and external operations for the ODNI, to include relationships with the White House, the Departments of Defense, State, Justice, and Homeland Security, the Central Intelligence Agency, and the Congress. He was also involved in the development of national intelligence centers, including NCTC and the National Counterproliferation Center, and their integration into the larger Intelligence Community. In addition, Mr. Leiter served as an intelligence and policy advisor to the Director and the Principal Deputy Director of National Intelligence. Prior to his service with the ODNI, Mr. Leiter served as the deputy general counsel and assistant director of the President's Commission on the Intelligence Capabilities of the United States Regarding Weapons of Mass Destruction (the "Robb-Silberman Commission"). While with the Robb-Silberman Commission, Mr. Leiter focused on reforms of the U.S. Intelligence Community, in particular the development of what is now the National Security Branch of the Federal Bureau of Investigation. From 2002 until 2005, he served with the Department of Justice as an Assistant

United States Attorney for the Eastern District of Virginia. At the Justice Department, Mr. Leiter prosecuted a variety of federal crimes, including narcotics offenses, organized crime and racketeering, capital murder, and money laundering. Immediately prior to his Justice Department service, he served as a law clerk to Associate Justice Stephen G. Breyer of the Supreme Court of the United States and to Chief Judge Michael Boudin of the U.S. Court of Appeals for the First Circuit. From 1991 until 1997, he served as a Naval Flight Officer flying EA-6B Prowlers in the U.S. Navy, participating in U.S., NATO, and United Nations operations in the former Yugoslavia and Iraq. Mr. Leiter received his J.D. from Harvard Law School, where he graduated magna cum laude and was president of the *Harvard Law Review*, and his B.A. from Columbia University.

ELIZABETH RINDSKOPF PARKER is dean emerita at the University of the Pacific, McGeorge School of Law. A noted expert on national security law and terrorism, Ms. Parker served 11 years in key federal government positions, most notably as general counsel for NSA; principal deputy legal adviser, Department of State; and general counsel for the Central Intelligence Agency. In private practice, she has advised clients on public policy and international trade issues, particularly in the areas of encryption and advanced technology. Ms. Parker began her career as a Reginald Heber Smith Fellow at Emory University School of Law and later served as the director, New Haven Legal Assistance Association, Inc. Early in her career, she was active in litigating civil rights and civil liberties matters, with two successful arguments before the U.S. Supreme Court while a cooperating attorney for the NAACP Legal Defense and Education Fund. Immediately before her arrival at McGeorge, Ms. Parker served as general counsel for the 26-campus University of Wisconsin System. She is a member of the Security Advisory Group of the DNI, the board of directors of the MITRE Corporation, the American Bar Foundation, and the Council on Foreign Relations, and she is a frequent speaker and lecturer. Her academic background includes teaching at Pacific McGeorge, Case Western Reserve Law School, and Cleveland-Marshall State School of Law. From 2006 to 2013, she held a presidential appointment to the Public Interest Declassification Board. Ms. Parker received her B.A. and J.D. from the University of Michigan.

PETER J. WEINBERGER has been a software engineer at Google, Inc., since 2003. After teaching mathematics at the University of Michigan, Ann Arbor, he moved to Bell Laboratories. At Bell Labs, he worked on Unix and did research on topics including operating systems, compilers, network file systems, and security. He then moved into research management, ending up as Information Sciences Research vice president, responsible for computer science research, math and statistics, and speech. His organization included productive new initiatives, one using all call detail to detect fraud and another doing applied software engineering research to support building software for the main electronic switching systems for central offices. After Lucent and AT&T split, he moved to Renaissance Technologies, a technical trading hedge fund, as head of technology, responsible for computing and security. He is a former member of the NRC's CSTB, current co-chair of an NRC committee on cybersecurity research, and served on several other NRC studies. He serves in a variety of other advisory roles related to science, technology, and national security. He has a Ph.D. in mathematics (number theory) from the University of California, Berkeley.

CONSULTANTS

M. ANTHONY FAINBERG became a research staff member at the Institute for Defense Analyses, where he focuses on risk assessment methodologies, countering nuclear terrorism, and nuclear non-prolieration issues, upon retiring from federal service after 20 years. At retirement, Dr. Fainberg was director of the Office of Transformational Research and Development of the Domestic Nuclear Detection Office of the Department of Homeland Security. Previously, he had been division chief at the Advanced Systems and Concepts Office, Defense Threat Reduction Agency, Department of Defense; before that, he directed the Office of Policy and Planning for Aviation Security in the Federal Aviation Administration. He also is a senior scientific advisor to the Pacific Basin Development Council, an organization comprising the governors of the U.S. Pacific island territories and Hawaii. He holds a Ph.D. in physics.

ALLAN FRIEDMAN is a research scientist at the Cyber Security Policy Research Institute (CSPRI) in the School of Engineering and Applied Sciences at George Washington University, where he works on cybersecurity policy. Wearing the hats of both a technologist and a policy scholar, his work spans computer science, public policy, and the social sciences, and has addressed a wide range of policy issues, from privacy to telecommunications. Dr. Friedman has over a decade of experience in cybersecurity research, with a particular focus on economic, market, and trade issues. He is the coauthor of *Cybersecurity and Cyberwar: What Everyone Needs to Know* (2014). Prior to joining CSPRI, Dr. Friedman was a fellow at the Brookings Institution and the research director for the Center for Technology Innovation. Before moving to Washington, D.C., he was a postdoctoral fellow at the Harvard University Computer Science Department, where he worked on cybersecurity policy, privacy-enhancing technologies, and the economics of information security. Dr. Friedman was also a fellow at the Kennedy School's Belfer Center for Science and International Affairs, where he worked on the Minerva Project for Cyber International Relations. He has also received fellowships from the Berkman Center for Internet and Society and the Harvard Program on Networked Governance. He has a degree in computer science from Swarthmore College and a Ph.D. in public policy from Harvard University.

ALEX GLIKSMAN is principal of AGI Consulting, LLC, a firm specializing in intelligence and other national security program development, congressional relations, acquisition, and management strategies. Mr. Gliksman has played a central role in the development and evaluation of analytic tools, and mission planning and operational support systems used by the Armed Services and U.S. intelligence for counterproliferation, counterterrorism, special operations, and arms control. He also has extensive experience in South and Southwest Asia and Pacific Rim regional security matters and has advised Fortune 500 companies on business opportunities in these regions. He served on the senior staffs of the House Intelligence Committee and Senate Foreign Relations Committee. Mr. Gliksman's clients have included the U.S. Congress, the Department of Defense, the Department of State, Lawrence Livermore National Laboratory, Sandia National Laboratories, Idaho National Laboratory, Northrop Grumman, the Analytic Sciences Corp., Booz-Allen & Hamilton, Science Applications International Corporation, the Boeing Company, and Computer Sciences Corporation. Mr. Gliksman has taught on the graduate faculty of the University of Southern California and at the University of Maryland. He studied at New York University and the University of Vienna and pursued doctoral studies in international relations at University College London.

STAFF

ALAN H. SHAW, *Study Director*, has been at the NRC as deputy director of the Air Force Studies Board since January 2014 and has had several previous assignments at the NRC. Educated as a physicist (Yale, 1974), he has worked in various capacities for all three branches of the federal government, including as the director for international security and space at the congressional Office of Technology Assessment. He has also worked for the Congressional Budget Office, the U.S. Arms Control and Disarmament Agency, the Institute for Defense Analyses, the Center for Naval Analyses, SRA International, the Defense Threat Reduction Agency, and (as a consultant) Lawrence Livermore National Laboratory and the federal judiciary.

JON EISENBERG is the director of NRC's CSTB where he oversees and directs studies and other activities related to computing, communications, and public policy. In 1995-1997 he was an AAAS Science, Engineering, and Diplomacy Fellow at the U.S. Agency for International Development, where he worked on technology transfer and information and telecommunications policy issues. He received his Ph.D. in experimental high-energy physics from the University of Washington and a B.S. in physics with honors from the University of Massachusetts, Amherst.

HERBERT S. LIN was chief scientist at NRC's CSTB until December 2014 where he had served as study director of major projects on public policy and information technology. These studies included a 1996 study on national cryptography policy (*Cryptography's Role in Securing the Information Society*), a 1991 study on the future of computer science (*Computing the Future*), a 1999 study of Defense Department systems for command, control, communications, computing, and intelligence (*Realizing the Potential of C4I: Fundamental Challenges*), a 2000 study on workforce issues in high-technology (*Building a Workforce for the Information Economy*), a 2002 study on protecting kids from Internet pornography and sexual exploitation (*Youth, Pornography, and the Internet*), a 2004 study on aspects of the FBI's information technology modernization program (*A Review of the FBI's Trilogy IT Modernization Program*), a 2005 study on electronic voting (*Asking the Right Questions About Electronic Voting*), a 2005 study on computational biology (*Catalyzing Inquiry at the Interface of Computing and Biology*), a 2007 study on privacy and information technology (*Engaging Privacy and Information Technology in a Digital Age*), a 2007 study on cybersecurity research (*Toward a Safer and More Secure Cyberspace*), and a 2008 study on health care information technology (*Computational Technology for Effective Health Care*). Prior to his NRC service, he was a professional staff member and staff scientist for the House Armed Services Committee (1986-1990), where his portfolio included defense policy and arms control issues. He received his doctorate in physics from MIT. Apart from his CSTB work, he is published in cognitive science, science education, biophysics, and arms control and defense policy. He also consults on K-12 math and science education.

ERIC WHITAKER is a senior program assistant at NRC's CSTB. Prior to joining the CSTB, he was a realtor with Long and Foster Real Estate, Inc., in the Washington, D.C., metropolitan area. Before that, he spent several years with the Public Broadcasting Service in Alexandria, Virginia, as an associate in the Corporate Support Department. He has a B.A. in communication from Hampton University.